# Basic Statistical Tools
# for Improving Quality

# Basic Statistical Tools
# for Improving Quality

**Chang W. Kang**

*Department of Industrial and Management Engineering*
*Hanyang University*
*South Korea*

**Paul H. Kvam**

*School of Industrial and Systems Engineering*
*Georgia Institute of Technology*
*Atlanta, GA*

A JOHN WILEY & SONS, INC., PUBLICATION

Published by John Wiley & Sons, Inc., Hoboken, New Jersey.

Published simultaneously in Canada.

For general information on our other products and services or for technical support, please contact our Customer Care Department within the United States at (800) 762-2974, outside the United States at (317) 572-3993 or fax (317) 572-4002.

Wiley also publishes its books in a variety of electronic formats. Some content that appears in print may not be available in electronic formats. For more information about Wiley products, visit our web site at www.wiley.com.

*Library of Congress Cataloging-in-Publication Data:*

Kang, Chang W. (Chang Wok), 1957–, author.
  Basic Tools and Techniques for Improving Quality / Chang W. Kang, Paul H. Kvam.
    p. cm
  Includes index.
    ISBN 978-0-470-88949-7 (pbk.)
1. Process control—Statistical methods. 2. Quality control—Statistical methods. 3. Acceptance sampling. I. Kvam, Paul H., 1962–, author. II. Title.
    TS156.8.K353 2011
    658.5'62—dc22                                                                                      2010042346

Printed in the United States of America.

10 9 8 7 6 5 4 3 2 1

*To Jinok, Philip, John*
*C.W.K.*

*To Lori*
*P.H.K.*

# CONTENTS

# PREFACE

*Basic Statistical Tools for Improving Quality* is intended for business people and non-business people alike who do not possess a great amount of mathematical training or technical knowledge about quality control, statistics, six sigma, or process control theory. Without relying on mathematical theorems, you will learn helpful quality management techniques through common sense explanation, graphical illustration and various examples of processes from different industries. Unlike most other books written about quality management, you will be *learning by doing*. Every time we introduce a new tool for improving quality, we will practice it on a real example. Programming skills are not necessary because we will implement a free menu-based and user-friendly software program that was designed with the basic tools and techniques for improving quality.

If you bought this book, you are probably familiar with some kind of production process. By *process*, we can mean a traditional manufacturing process such as goes on in a Hyundai Motors automobile assembly plant in Korea, but it can also refer to a supply chain that links a Maple tree seedling from an Oregon plant nursery to a Baltimore garden store. In a service industry, a process can be the stages necessary to provide the customer at McDonald's with an Egg McMuffin.

All organizations have a critical process and strive hard for improving quality of products and/or services through that process. At the end of each process, there always exists some bit of variation in output. For example, in the chocolate chip cookie manufacturing process, the number of chocolate chips in each cookie varies. This is not an exciting prospect for the customer who opens the cookie box with clear idea of what they want and expect in a chocolate chip cookie: not too many chips, and certainly not too few. But some variation between cookies is inevitable, and these variations are difficult to see during the baking and packaging process.

In order to improve quality, it makes sense to reduce this item-to-item variability when we can. But to know about the variability in the process, we have to measure things, and then figure out how to assess this variability by analyzing the resulting data. This can get complicated, even if we are baking chocolate chip cookies, where it is impractical to count the chips in every baked cookie. This is where statistical process control takes over.

*Statistical process control* refers to the application of statistical tools to measure things in the process in order to detect whether a change has taken place. In the examples we have discussed, the process aims for consistency, so any change in the process output is usually a bad thing. This book describes and illustrates statistical tools to detect changes in a process, and it also shows how to implement the idea of continuous improvement into process control.

The tools you can implement are both graphical and analytical. Graphical procedures (charts, diagrams) serve as effective tools to communicate intricate details about a complex or dynamic process without having to go into statistical or mathematical detail. Graphical tools are helpful, but not always convincing. Analytical tools, on the other hand, can be more powerful but are more challenging to learn. While we spare the reader of any unnecessary detail about the mathematical machinery of the analytical tools, we certainly do not promote an overly simple *"push that button"* mentality in this book. That won't work for the unique problems encountered when applying these tools and techniques to your workplace. The analytical tools need a better level of understanding.

To be an effective process manager, you don't have to be an expert in statistics. You do need to be knowledgeable about the process you are working on, and you need to be determined enough to learn how these simple tools can be used to understand and control this process. This is the kind of expert that we hope you will become after you have finished reading this book. To avoid unnecessary statistical formulas, we will focus on concepts and show you how to use the free eZ SPC software to understand how data is analyzed. We will learn through examples.

Here are some key skills you should pick up after reading *Basic Statistical Tools for Improving Quality*:

1. You will perceive and acknowledge that there is always some kind of problem in the process.

2. You will know what kind of problem solving tools or techniques are required to overcome your obstacles.

3. You will know how to collect appropriate data that will make the problem solving as easy as possible.

4. You will know which tools to use and how to use menu commands from the eZ SPC software to complete your data analysis.

5. You will know how to interpret the eZ SPC results to explain how the problem is solved.

## Outline of Chapters

- Chapter 1: In the first chapter, you will learn how to frame a typical business task or industrial procedure as the kind of process that is appropriate for quality improvement. Examples will be used to illustrate processes we encounter in daily life and we outline how tools and techniques for quality improvement can be learned and exploited.

- Chapter 2: Here, we introduce you to eZ SPC, the free and easy-to-use software that will help you analyze and solve quality management problems in this book and at your job. If you are already comfortable programming with other software packages that are capable of running all the necessary statistical process control procedures listed in this book, you can take advantage of your programming expertise without using the eZ SPC program. However, we devote parts of this book to helping the reader complete quality management tasks using the simple menu-based commands of eZ SPC. Chapter 2 shows you how charts and graphs can be created to help summarize process performance and describe how efficiently a process is working. We include bar graphs, box plots, cause-and-effect diagrams, histograms, Pareto charts, pie graphs, radar charts and scatter plots.

- Chapter 3: In this chapter, we will go over the key features in a set of data. To know which values appear in the data set and how frequently they are found at different values tells us about the data's *distribution*. You will learn about statistics that describe the spread of the data as well as where the middle or end-points of the data set are. Using helpful tools in eZ SPC, we will learn how to summarize data and find out about how it is distributed without having to use graphs.

- Chapter 4: As a companion chapter to Chapter 3, here we will present you with an overview of the analytical tools you will rely on in the analysis of process data. Two key concepts in data analysis are confidence intervals and statistical tests of hypotheses. These procedures not only provide a summary and prediction of the process from the features of the data, but they also provide a measure of our uncertainty regarding those predictions. The ability to communicate the uncertainty in statistical results will make our results more objective, informative and scientific.

- Chapter 5: Here we introduce you to a primary method for evaluating and monitoring a process: the control chart. Control charts are simple and varied graphical tools that can be easily generated using the eZ SPC software, and the generated chart helps us to determine if the process is running correctly. The various kinds of charts represent the various goals of the process manager as well as the different kinds of measurements from that process that we can scrape together for data analysis.

- Chapter 6: The control charts in Chapter 5 are constructed to detect the first moment when a process breaks down and goes out of control. In the following chapter, we move from these basic control charts to more advanced charts that are constructed to detect the incremental changes made by a process that is slowly degrading or getting out of control. Also in this chapter, you will learn how charts and statistics can be applied to assess process capability once the control charts have shown that the process is running as planned.

- Chapter 7: Here we will show how you can create process improvement by investigating potential factors (in the environment or in the process itself) that affect the process and/or the process output. Some of these factors are readily identifiable, but there might be other causes and factors that are not obvious. Moreover, these new factors we consider not only affect the process output, but they may affect each other. That can really challenge our ability to understand the process the way we want. To overcome this challenge, statistical methods such as correlation analysis, regression, analysis of variance and factorial design can be implemented using eZ SPC.

- At the end of the book, we include a glossary of terms that are used in the book, and we also include some terms not found elsewhere in the book. These terms generally fall outside the scope of statistical process control, but you might come across them in your work on process improvement.

## Examples, Exercises and Test Questions

Examples are featured throughout the book and easy to find. Just look for a gray box followed by the example text. Here is an example from the first chapter:

Example 1-8: General Hospital Emergency Center

Hospitals must always be at the forefront of process improvement. Patient processing at an emergency reception has evolved continuously over.... The example always ends with three dots:

• • •

Exercises are made to challenge the reader to think about a problem, and they might require the reader to use eZ SPC for solving it. Because this is a self-teaching textbook, we try to guide you through the steps of the exercises, and discuss the outcomes you should be able to achieve.

At the end of each chapter, we include a section titled *Test Your Knowledge* that contains 10 quiz questions. These will allow you to quickly evaluate your learning as you go along, and generally don't take as long to finish as the exercises. Solutions to the quiz questions are found in the last section, title *End Matter*. In that section, we have also constructed a Final Exam consisting of 100 multiple choice questions that cover all the chapters of the book.

By the time you finish the book and the exercises, you can be confident that you have mastered the basics of process control and improvement. Then, you will be ready to take what you have learned and practice it on all of the processes that inspired you to get this book in the first place.

## Why We Wrote This Book

Before this book was imagined, one of the authors (Dr. Kang) developed and improved software tools for use in quality management training workshops. Participants who attended the workshops included mid-level managers, assembly line foremen, and a few people in upper management. The audience was attending the seminar for one reason: to obtain the minimum hours of training they needed to help them achieve ISO 9000 accreditation and certification (ISO stands for International Organization for Standardization). As a result, participants were usually satisfied to passively sit through workshop seminars, which (let's face it) can tend to be dull. The initial results were not positive.

To improve learning in later workshops, class participation was mandated through use new software tools made available to the class. Along with lessons in quality management, participants made graphs and analyzed process data with a click of the button. As a result, the audience became more involved in learning how statistical process control tools could be used in quality management without having to learn about complicated statistical formulas or detailed computer algorithms. It was all available to them with a push of the button, and participants were excited to use these newly learned techniques on quality management problems.

The software was further refined and improved. We listened to what process managers told us about which tools and techniques were most helpful to them in their day to day work. It was easy to gather the most important ones to put in this book, but it was not so easy to decide what to leave out. For example, multivariate control charts are an important foundation in advanced process control monitoring, but seemed unnecessary for learning the basics of process control. So we grudgingly left it out, along with some other highly advanced topics in statistical process control so we could focus on the important stuff.

**Free Software**

With the software in mind, we will show the reader how to make statistical graphs and analytical tables for quality management problems. With every new lesson, there will be a computer tool to help you solve problems. Exercises and quiz questions will also include a dose of pain-free data analysis using the provided software. The software program eZ SPC is all you need, and you can download it for free at

$$http://www.hanyang.ac.kr/english/ezspc.html$$

Along with the program, you should download the Applications Folder, which contains all of the data sets introduced throughout the book. These data sets are used both in examples in the textbook and quiz questions at the end of every chapter. Of course, you can also use eZ SPC for your own statistical analysis and process control once you are finished reading this book. The spreadsheet structure of eZ SPC allows an easy transfer of data from text files and Microsoft Excel© files.

## Acknowledgments

Along with our families, we are grateful to our students, colleagues and the editors. From Wiley, we thank Jacqueline Palmieri, Melissa Yanuzzi, Amy Hendrickson and Steve Quigley.

We thank Dr. Bae-Jin Lee and Dr. Sung-Bo Sim for their contribution to the development of eZ SPC software, Dr. Jae-Won Baik, Jong-Min Oh, and Eui-Pyo Hong for their contributions to the development of eZ SPC 2.0. We would also like to thank Dr. Hae-Woon Kang for his special contribution to develop and improve eZ SPC 2.0 English version for this book. We also thank current and former graduate students of the Statistical Engineering for Advanced Quality Laboratory, Hanyang University who played supporting roles as we improve eZ SPC software and as we prepare this book. We would also like to express our appreciation to the many students and Dr. Kyung Jong Lee who have used eZ SPC 2.0 Korean version and who have made useful suggestions for improvement of the software.

We have benefited from Hanyang University colleagues who encouraged us working together at Georgia Tech. Thanks go to colleagues at Georgia Tech who made this cooperation possible, including Chip White, Lorraine Shaw and Michael Thelen.

CHANG W. KANG
*Department of Industrial & Management Engineering*
*Hanyang University*
*South Korea*

PAUL H. KVAM
*H. Milton Stewart School of Industrial and System Engineering*
*Georgia Institute of Technology*
*United States*

# CHAPTER 1

# THE IMPORTANCE OF QUALITY IMPROVEMENT

## 1.1 INTRODUCTION

Think about how rapidly things change in today's world. As the years have gone by, many of our daily chores and activities have increased dramatically in speed. With on-line bill paying, internet commerce and social networking available on mobile devices like smartphones, in one hour we can finish a set of tasks that once took a full day to accomplish.

The business world changes rapidly, too. Once a company has brought a new product to market, they probably have only a short time to celebrate this accomplishment before competitors introduce an improved version or a new and unexpected alternative that customers may soon prefer. Because as technology changes, customers change, too. It's easier for consumers to collect information about new products so they can make educated choices. Today's consumers have higher expectations and less tolerance of products that don't keep up.

In this way, the business world emulates our daily life. Instead of using all of our saved time to enjoy the additional hours of leisure that new technologies have afforded us, we immediately reinvest our time in new tasks that help us

*Basic Statistical Tools for Improving Quality.* By Chang W. Kang and Paul H. Kvam    **1**
Copyright © 2011 John Wiley & Sons, Inc.

produce even more so we can keep up with our colleagues, adversaries and neighbors. And it's even tougher for businesses these days, because globalization of markets has brought together competitors from every part of the world, not just from across town.

As the world melds into one enormous trading village, increased competition demands faster increases in product output and quality. More than ever, weaker companies will be driven out of the market without protection of domestic loyalties or the fleeting fortunes derived from the economic benefits of assumed scarcity. Consider the impact of Japanese car manufacturing on Ford and General Motors in the 1970s and 1980s. When automobile purchasers in the United States were more limited in their choice of new automobiles, the car producers took advantage of the scarcity of customer resources and did not aggressively pursue quality improvements until they lost the scarcity advantage to Japanese companies such as Toyota, Honda and Mitsubishi.

Starting in the 1970s, the confidence and reputation enjoyed by General Motors and other domestic auto makers slowly eroded up through the 1990s, when customers in the United States tacitly assumed a new automobile from Toyota, Nissan or Honda would deliver higher quality at a better price than the domestic alternatives. In the coming years, auto makers with a reputation for high quality products will have to work harder to retain their customers' confidence because competitors will catch up more quickly than it took American automobile manufacturers to react to the flourishing Japanese automobile industry.

In 1990, Japan became the world's top producer of automobiles. In 2008, Toyota surpassed General Motors as the largest auto maker in the world. Late in 2009, however, Toyota started to face millions of potential car recalls due to problems with acceleration. All of a sudden, Toyota's reputation for long-term quality is being severely tested. Customers are more knowledgeable, so manufacturers and service providers that fail to adapt to meet the needs, expectations, and requirements of customers will lose market share quickly. By early 2010, for example, Toyota lost a third of its U.S. market share in less than a month after the car recalls were announced. Other Japanese auto makers have experienced reduced market share as well, in part due to recent competition from Korea.

With their increased access to knowledge, consumers have demanded higher quality products and more responsive services. With the ruthlessness of global competition, companies are frantically adopting various strategies for improving quality and reducing cost. These strategies include statistical quality control, total quality management, Six Sigma and Lean Six Sigma. Statistical quality control is the chief way many manufacturers realize quality improvement, and this includes sampling plans, experimental design, variation reduction, process capability analysis, process improvement plans, reliability analysis and statistical process control.

Of all these various components, statistical process control (SPC) is the focal point of quality improvement. The techniques we learn in this book will make up the toolbox you need to analyze and explain problems with process management. It requires some *statistical thinking* where we frame our problems in terms of processes, consider how these processes are interconnected, and how reducing the variation in these processes will be a key to our success as a quality manager. Once these problems are addressed, we can focus on process improvement. In the next section, we explain just what process control means in business and manufacturing, and how it differs from other quality control management tools.

## 1.2  WHAT IS STATISTICAL PROCESS CONTROL?

In general terms, statistical process control can be defined as a method which monitors, controls, and ideally improves a process through statistical analysis. SPC includes measuring the process, reducing variability in the process to make it produce a consistent output, monitoring the process, and improving the process in terms of a *target value* of the output. By monitoring the process, diagnostic techniques are applied to identify different causes of variation in the process, eliminate the *assignable causes* of variation, and continuously improve the process in order to reduce variability or production cost. An assignable cause might be some unnatural or unwanted pattern observed in the process. This is in contrast to a *common cause*, which more likely is treated as random, unavoidable *noise* or *natural variability* in the process. This chance-cause variation can be as innocuous as a minor measurement error.

Statistical process control uses statistics to analyze the variation in processes. This represents the more technical side of quality improvement, and requires the most effort to understand. In this book, we guide the reader around (and away from) most of the technical details required in the statistical analysis of process data, but we emphasize the importance of the statistical issues and how statistical procedures are used to solve quality management problems. Using statistics, SPC tries to keep the process average at a target value and to reduce the unpredictability of the process. These techniques can be applied to a wide variety of industries, to maintain the consistency of a manufacturing process, a service process, or any of the unique processes in today's business world.

## 1.3  THE BIRTH OF QUALITY CONTROL

The birth of quality control started around Chicago back in 1924, at the same time when prohibition helped make the city notorious for bootlegging and gangsters. The most notorious gangster of this time was Al Capone. By 1920, Capone was just making a name for himself among the Chicago crime syndicate, but in a few years, he was well on his way to becoming America's most famous gangster of all time. At the height of his career, Capone's crime group ran an impressive bootlegging operation, controlled speakeasies, gambling houses, brothels, race tracks, and nightclubs that brought in income of nearly 100 million dollars per year. When the operation moved to the Chicago suburb of Cicero in early 1924, the city of Cicero became synonymous with mobster life and crooked politics.

(a)                              (b)

**Figure 1.1**    Industry innovators of Chicago in the 1920s: (a) Al Capone (1899 - 1947) and (b) Walter Shewhart (1891 – 1967).

But Capone didn't have the biggest operation in town. The Western Electric Company, also based in Cicero, was producing 90% of the nation's telephone equipment, and brought in over 300 million dollars per year. The company's Hawthorne Works plant had become a world famous research center, developing the high-vacuum tube, the condenser microphone, and radio systems for airplanes. One 1923 study that originally set out to show how better lighting increased worker productivity later became famous for its discovery of a short-term improvement caused by observing worker performance, which psychologists refer to as the "Hawthorne effect".

Less famously at the time, Western Electric pioneered the development of inspection policies to assure specification and quality standards in manufactured products. When an engineer and statistician named Walter Shewhart joined the Inspection Engineering Department at Hawthorne in 1918, industrial quality was limited to merely inspecting products and removing defective ones. Shewhart changed all that on May 16, 1924, when he presented his boss, George Edwards, a memo that would soon change the way we perceived quality control in manufacturing. Edwards remarked

> "Dr. Shewhart prepared a little memorandum only about a page in length. About a third of that page was given over to a simple diagram which we would all recognize today as a schematic control chart. That diagram, and the short text which preceded and followed it, set forth all of the essential principles and considerations which are involved in what we know today as process quality control."

Shewhart's ensuing research built on this genesis of statistical process control, with many of his early results published in his 1931 book, *Economic Control of Quality of Manufactured Product.*

It would be exciting to think that Walter Shewhart and Al Capone were somehow intertwined in 1924 Cicero, but this is highly unlikely. It seems more probable that Chicago's bootlegging operation lacked statistical process control of any kind. The closest tie we can make is from what we know of Shewhart's protégé, Joseph Juran, who occasionally visited a couple of Capone's casinos across the street from the Hawthorne Works. After spending some time in Capone's establishment after hours, Juran noticed that one roulette wheel operator worked "like a robot", making the operation of his wheel amenable to statistical analysis and prediction. His expertise enabled him to win one hundred dollars, which at the time was several weeks' pay.

Shewhart deduced that while variation exists in every type of process, some variation could be blamed on an assignable cause (so it is not completely random) but other sources of variation could not. The key was to focus on the factors that are associated with assignable cause variation in order to remove as much of that variation as possible. Inherent or natural variation, on the other hand, has only chance cause (or common cause) and can only be tolerated.

Along with Juran and his other famous protégé, Edward Deming, Shewhart revolutionized the science of manufacturing processes. Their research was eventually disseminated throughout the world, but ironically did not catch on quickly at Western Electric. In 1925, Western Electric joined with AT&T, the co-owner of Bell Labs, and much of its research was consolidated and moved to Bell Labs research centers in New Jersey. The Hawthorne Plant was greatly downsized by the Great Depression, and Juran noted by the time he left in 1941, "you could walk through this plant, the seed bed of the quality revolution, without seeing any control charts".

Deming is famous for teaching American manufacturers about quality and management, but his work was first adopted by the Japanese Union of Scientists and Engineers, whose members had already studied Shewhart's techniques. Starting in the 1950s, Deming trained a generation of Japanese scientists, engineers and managers in quality improvement. In the 1980s, many believed this training was one factor that helped the Japanese automobile industry leap-frog over its overseas competitors and allowed Japanese electronics manufacturers gain a large market niche in world trade. Later, American manufacturers such as Ford Motor Company also recognized the value of Deming's techniques, although Ford waited until 1981 to hire him for quality consultation.

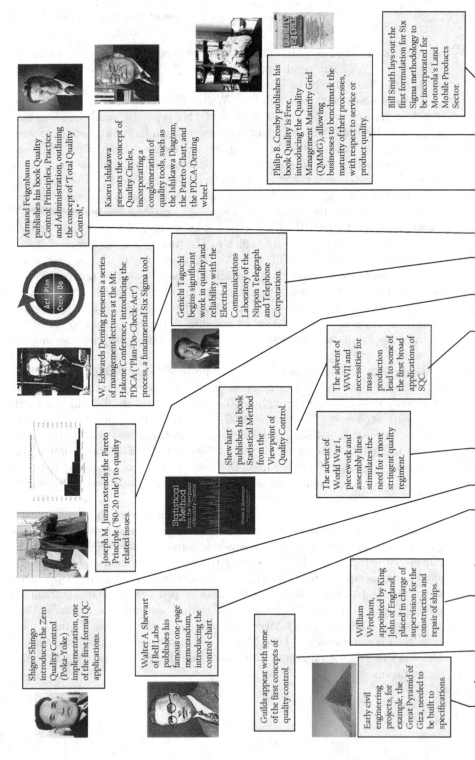

Armand Feigenbaum publishes his book Quality Control: Principles, Practice, and Administration, outlining the concept of "Total Quality Control."

Kaoru Ishikawa presents the concept of Quality Circles, incorporating a conglomeration of quality tools, such as the Ishikawa Diagram, the Pareto Chart, and the PDCA-Deming wheel.

Philip B. Crosby publishes his book Quality is Free, introducing the Quality Management Maturity Grid (QMMG), allowing businesses to benchmark the maturity of their processes, with respect to service or product quality.

Bill Smith lays out the first formulation for Six Sigma methodology to be incorporated for Motorola's Land Mobile Products Sector.

W. Edwards Deming presents a series of management lectures at the Mt. Hakone Conference, introducing the PDCA ("Plan-Do-Check-Act") process, a fundamental Six Sigma tool.

Genichi Taguchi begins significant work in quality and reliability with the Electrical Communications Laboratory of the Nippon Telegraph and Telephone Corporation

The advent of WWII and necessities for mass production lead to some of the first broad applications of SQC.

Joseph M. Juran extends the Pareto Principle ('80-20 rule') to quality related issues.

Shewhart publishes his book Statistical Method from the Viewpoint of Quality Control.

The advent of World War I, piecework and assembly lines stimulates the need for a more stringent quality regiment

Shigeo Shingo introduces the Zero Quality Control (Poka-Yoke) implementation, one of the first formal QC applications.

Walter A. Shewhart of Bell Labs publishes his famous one-page memorandum, introducing the control chart.

Guilds appear with some of the first concepts of quality control

William Wrotham, appointed by King John of England, placed in charge of supervision for the construction and repair of ships.

Early civil engineering projects, for example, the Great Pyramid of Giza, needed to be built to specifications.

**Figure 1.2** A timeline of the history of statistical process control in the 20th Century

## 1.4 WHAT IS A PROCESS?

A process is defined as a series of operations performed in the making of a product. More simply, a process is the transformation of input variables to a final result that might be a product or service. We will often refer to the process result as a product or an output - see Figure 1.3. In a manufacturing process, the input variables can be anything from raw materials, machines or equipment, operators and operator actions, environmental conditions, solicited information and working methods. Some factors are clearly controllable, such as an operator's action in the process. Uncontrollable factors include inputs that clearly affect the output but are not under our control, such as the effect of weather on a house construction process.

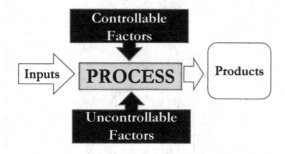

**Figure 1.3**  Simple diagram of a process affected by controllable factors and and uncontrollable factors

The variation in the inputs will propagate through the process and result in variability in the output. In general, process managers will want a predictable process that produces identical items consistently, so any variability in the output is bad news. We know that perfect consistency in the output is an unobtainable ideal, so we have to deal with the output variability by understanding it, characterizing it, and ultimately reducing it. The output variability is created by assignable causes or chance causes or both. The variability created by assignable causes is usually unpredictable, but it is explainable after it has been observed. Some examples of assignable causes are machine malfunction, low quality batch of raw materials, new operator, and so on. Chance causes produce random variation in the behavior of output measurements. This random variation created by chance causes is consistent and predictable, but it is unexplainable. Some examples of chance causes are vibration in the processes, poor design, noise, computer response time, and so on. Assignable causes contribute significantly to the total output variability.

Due to assignable and chance causes, output items that are distributed like this bell-shaped curve are characterized statistically using a Normal distribution (see Section 4 of Chapter 2: Statistical Distributions). In Figure 1.4, the target value is a well known ideal value and LSL  (lower specification limit) and USL (upper specification limit) represent what outputs can be tolerated in the market. The target value, LSL, and USL are set by product standards or by the engineers. The LSL and USL are used to determine the products are acceptable. For example, the producer of bottled water can set the target value for the content of a bottle to be 500 ml and he can also specify that any quantity between 495 ml and 505 ml is acceptable. Here, the target value is 500 ml, LSL is 495 ml, and USL is 505 ml. Even though the output variability is understood to exist, process engineers must try to produce each product to be between LSL and USL.

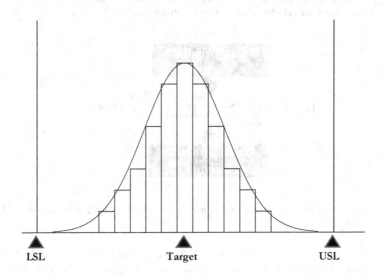

**Figure 1.4**   Distribution of output measurements for a process

## 1.5   EXAMPLES OF PROCESSES FROM DAILY LIFE

In this book, we consider a large variety of process examples that are subject to statistical process control . This will provide readers with a wide array of applications and allow each reader to identify with some real-life applications when learning how to improve process quality. For example, one reader might closely relate their own work experiences with examples in semiconductor manufacturing, while another reader associates more with the process of shipping manufactured goods from China to a series of distribution centers in the United States.

## Example 1-1: Frying an Egg

Suppose that we are at home, frying an egg for breakfast. In this process, the input variables are the cook, a pan, a gas or electric range, an egg, cooking oil, salt, an egg turner, and the recipe that provides you with instruction on how to fry an egg. There are many factors involved in the process such as cooking temperature, cooking time, amount of cooking oil on the pan, amount of salt, and so on. Among them, we can control the cooking time, the amount of cooking oil, and the amount of salt. These are controllable factors. If we want to make a consistent output in our fried egg process, we need to be consistent with the levels of the controllable input factors. However, the cooking temperature and uniformity of the input egg can't be controlled exactly. In this process, they are uncontrollable factors. Due to the uncontrollable factors, we experience slightly different results every time we cook a fried egg. This represents the variation in the egg-frying process.

• • •

## Example 1-2: Personnel Hiring

Suppose a computer software firm wants to hire an electronics engineer for software and hardware design. There are numerous steps in the hiring process, and uncontrollable factors that affect the process goal include the quality and quantity of the candidate pool as well as the demand from competing companies. Like many service processes, in this example we keep track of a person throughout the process, and give less thought to changeable inputs of machine parameters and raw materials. The process might be condensed into eight steps:

1. Advertise for the position, possibly including proactive calls to potential candidates who are working for other companies. The resources used in this step represent a controllable factor, given some firms spend much more money for advertising than other firms.

2. Form hiring group to judge resumes and select appropriate candidates.

3. After discussion and debate, a candidate is selected and then called by the department head and the human resources (HR) group. If turned down right away, the process goes back to step 2.

4. Negotiate with candidate to select date for interview, and set up travel and interview sessions.

5. Interview candidate. The hiring company has several controllable factors in the interview process, from the scheduling, the quality of transportation, lodging and restaurants used during the interview.

6. Hiring group discusses candidate and decides whether or not to tender an offer (if not, it's back to step 2). The offer will be formalized by the department head and the HR group.

7. The department head and HR group will negotiate with the candidate, if necessary, to entice the candidate to accept the offer (or it's back to 2 again). If the company is flexible, controllable factors can be used to optimize the contract offer. The prejudices, needs and preferences of the candidate are uncontrollable factors, and they play a crucial role in whether the offer is accepted.

8. The candidate signs an employment contract with the company. This is a less mechanized process, and it depends on subjectivity, group consensus and possibly unknowable hiring resources. Although there are numerous controllable factors involved in such a hiring process, factor levels may be ambiguous, especially to someone not on the hiring committee.

• • •

## Example 1-3: Renting a Car

Suppose that we are picking up a rental car at the airport. The process includes the following nine steps:

1. Fill out the contract document at the agency desk.

2. Select a car from available car list.

3. Decide whether to purchase the insurance plan.

4. Select option items such as all-wheel drive, a GPS navigator or a child safety seat.

5. Register a credit card with the rental agent.

6. Receive the car keys and the signed rental agreement.

7. Go to the car pick-up area (by walking or using shuttle bus).

8. Inspect the car for scratches and dents.

9. Start engine and drive the car out of the airport vicinity.

In this case the inputs associated with service labor are somewhat clear. What might be hidden, however, are the numerous sub-processes that must work in order to guarantee the right car is available to the customer at the right time. Along with the quality of the rental car, the process output might be characterized by service quality, such as the time when the customer enters the queue at the rental agency desk to the time the customer starts the rental car. The controllable factors include rental car selection, the number of rental agents working at the service counter and the training, competence and courteousness exuded by the agents. Uncontrollable factors can include unforseen customer demand affected by airline delays, delays caused by difficult customers, and the effects of nearby car rental agency competitors on customer demand.

· · ·

## Example 1-4: Filling a Drug Prescription

Suppose that a pharmacist takes a drug prescription from a customer and fills the order for the customer. We consider the following nine steps in the process:

1. Pharmacist receives doctor's prescription for medicine at front counter.

2. Pharmacist verifies customer prescription.

3. Pharmacist verifies insurance information

4. Pharmacist logs prescription into queue for retrieval.

5. Pills are obtained from shelves and counted to match customer order.

6. Pills are inserted into pill bottle, which is labelled according to the prescription.

7. Filled prescription is taken to check out to be picked up by customer.

8. Upon arrival at check out, customer is offered advisement on instructions or side effects with regard to the prescription.

9. Customer purchases drugs.

If the time needed to fill the prescription is the main process output, then the customer's actions might represent the most critical uncontrollable factor, since they might ask for the prescription right away. Controllable factors that hinder the pharmacists speed can be an outdated cash register, an inefficient referral system, or inadequate inventory for popular medicines.

• • •

## Example 1-5: Juice Manufacturing

Some farms, groves and vineyards have resources to harvest, process and package food products without the help of middlemen or large agri-businesses. For a berry grower, there are harvesting, pressing and bottling machines needed to make the process efficient, but given such machines are available, we might consider machine efficiency and reliability as uncontrollable factors. The process of turning berries into packaged juice products can be summarized into six basic steps:

1. The harvest and sorting of berries. There are numerous uncontrollable factors relating to weather, climate and soil quality, and controllable factors might include allotted growing time and number of field workers hired to help harvest the fruit.

2. Wash berries at juicing facility. Uncontrollable factors: amount of dirt clods, weeds and stones to remove when sorting berries. Controllable factor: amount of water and human labor used in washing berries.

3. Juice berries to puree via cold pressing through screening that removes seeds and stems. The screen filtering is a controllable factor.

4. Samples are examined to measure nutrients and detect contaminants such as pesticides. Factors are controlled through setting measurement tolerances, but potential measurement error might serve as an uncontrollable factor.

5. Blend juices and combine needed additives (factors same as step 4).

6. Bottling juices can include capping, labeling and boxing.

• • •

## Example 1-6: Abused or Neglected Children

Not all industry processes are about making things or selling things customers. The Division of Family and Children Services (DFCS) in the state of Georgia is the part of the Department of Human Resources (DHR) that investigates child abuse. DFCS finds foster homes for abused and neglected children and, at the same time, help their impoverished parents get back on their feet with counselling and job training. Like many other service providers, the inputs (children, families) provide a lot of uncontrollable variation to the process.

1. DFCS is alerted (perhaps by a nurse or school teacher) about a case of child abuse or neglect and writes up a report. The reporting network is an important controllable factor that can vary greatly between different communities.

2. The report is screened for potential investigation. If the case meets the criteria for abuse or neglect, DFCS sends investigators to the child's home to check on the health of the child. They inspect living conditions, talk with the parents and interview people involved with this case.

3. A decision is made about the case of abuse or neglect, and a wide range of outcomes are possible, from a formal warning for the parents to obtaining a court order to remove the child from the home. In some cases, the police will be involved, representing another potential source of uncontrollable process variation.

4. In cases where abuse charges were substantiated but the child is not in eminent danger and remains in the home, case managers visit the family regularly and provide services that might include counselling, drug abuse treatment, referrals for employment or child care.

5. If the child has to be removed from the home and not into the care of a family relative, a court order might be used to terminate parental rights, at which time the child is sent to foster care and officially becomes a ward of the state.

This is an intricate process with numerous uncontrollable factors related to the abuse, the physical and mental state of the parents and the family's resources. In both steps 3 and 4, the case manager decisions are based on a finite but complex set of controllable factors requiring the expert decision making of a trained social worker.

· · ·

## Example 1-7: Amazon.com

Amazon.com is a premier on-line sales site for merchandise such as books, dvds and music cds. One crucial part to Amazon's success is maintaining a fast, simple and easy process for the customer to select and purchase merchandise. To ensure the process runs smoothly, Amazon uses six basic steps to get the customer to complete a purchase on line:

1. After the customer clicks on a product to learn of product details, a bright yellow button that reads [Add to Cart] is displayed prominently next to a picture of the product.

2. Once the [Add to Cart] button is pressed, Amazon leaves several opportunities to shop some more (and includes suggested items based on what is already in the shopping cart) but the bright yellow [Proceed to Checkout] button is the most conspicuous one on the page.

3. If the customer has not yet signed in to their personal account (or signed up for one), they are required to do so before proceeding to the next stage of the process. For a computer user that allows Amazon's cookies, this can be as simple as typing in a password and clicking [Continue].

4. The customer selects a mailing address or types in an address if this is the first order. By storing addresses along with other customer information for use in expediting the order, Amazon can take advantage of the customer's potential impulse buying habits.

5. One click will determine the shipping method from a short list of options, and buttons that encourage the customer to continue with the order are always featured with bright yellow buttons.

6. If credit card information was previously used and stored by Amazon, the next page will finish the customer's order with a button for [Pay With Existing Card] and another bright yellow [Continue] button.

Without having to type a single letter beyond the customer's account password, merchandise can be selected and processed with only eight clicks of a mouse. By making the process so simple, Amazon has removed chances for process variability from uncontrollable factors related to the internet connection, user distractions and the customer's latent hesitancy in spending money so quickly. Even controllable factors are restricted in order to ensure the process runs smoothly. These factors include giving the customer buying suggestions based on what is in their shopping cart.

• • •

## Example 1-8: General Hospital Emergency Center

Hospitals must always be at the forefront of process improvement. Patient processing at an emergency reception has evolved continuously over the past decades. Let's consider a very simplified outline of the steps used in processing a patient into an emergency treatment center.

1. Patient arrives at registration via any possible mode, including ambulance, walk-in, or appointment.

2. Patient (or care taker) receives registration form, fills them out and returns them to the receptionist upon completion.

3. A clerk at the hospital registration desk enters the data from the completed forms. Data are verified to determine if the patient is able to receive medical treatment.

4. A triage nurse assesses and ranks patients based on the severity of the patient's illness or injury. The emergency severity index (ESI) for triage acuity is used to prioritize patients on a scale of 1-5:

   1 Resuscitation (heart attack or stroke)

   2 Emergent (e.g., critical fluid build-up)

   3 Urgent (e.g., excessive vomiting)

   4 Nonurgent (e.g., broken finger)

   5 Referred

5. Registered nurse re-assesses patient illness, provides minor standardized treatment to assessed illness, and prepares the patient for the physician's diagnosis.

6. Once the patient is treated by the nurse, a physician comes in to assess the patient.

7. Radiology and lab testing help diagnose patient illness. Lab testing occurs at the hospital's medical laboratory. Results from patient blood and urine tests are called in to physicians when they become available.

8. Once laboratory results are returned, the emergency department (ED) physician must make a diagnosis and decide whether to admit the patient into the intensive care unit (ICU) or discharge the patient from the ED. To reach a diagnosis and a decision, the ED physician contacts one or more medical consultants. Consultants can be the patient's family

doctor or a physician who specializes in a certain medical field. Once the consultant is reached and a decision finalized, the patient can be discharged or admitted.

9. Patients who are admitted to the hospital or discharged from the ED must have paper work filled out before they leave.

• • •

## 1.6   IMPLEMENTING THE TOOLS AND TECHNIQUES

The idea of using statistical process control to improve quality is usually supported enthusiastically by business and project managers. The implementation of statistical procedures, however, is another matter. This kind of quality improvement requires total commitment from top management and involvement of all employees. The participation and dedication showed by top management is essential to the success of process control. To involve workers effectively, sufficient attention must be made to employee training. Even though less-than-fully trained employees can successfully administer a well documented process, they will tend lack enthusiasm and sufficient understanding of the importance of each component in the process.

First, companies need to identify and focus on the key processes for implementation of statistical process control. They must begin on a single key process and study the procedures and results carefully. After confirming that the controls are working and are helping quality improvement and cost reduction, they can then apply these tools and techniques to other key processes.

Training requires more than providing workers with the required knowledge of process control techniques. Without training the top managers, the implementation of quality improvement is hindered by management's lack of vision and imagination regarding how these techniques can be employed to help in process improvement. In a broad sense, process improvement must be implemented repeatedly, and not just as a reaction to evidence of decreasing product quality. This leads us to the concept of *continuous process improvement*.

## 1.7   CONTINUOUS PROCESS IMPROVEMENT

Continuous process improvement is essential for improving quality and productivity. Knowing what to measure is a fundamental issue in implementing statistical process control. This requires knowledge about the product and its potential utility, as well as the qualities of the product that are measurable. But this is just one of several important rules we need. The following eleven steps outline the course of action for achieving continuous process improvement.

1. **Define the problem in the process**. This sounds simple, but there might be several candidate problems to consider. We will show you how to use the Pareto Principle to help you define the problem. Even if the process appears to be running smoothly, there may be some hidden problems or potential points for improvement that can be treated. In this step, it is best to work with a team, brainstorming together to collect candidate problem areas for possible improvement.

2. **Prioritize potential problem areas**. Select the primary candidate problem by using an appropriate emphasis on effectiveness, feasibility, urgency, capability, and conformance to business goal.

3. **Select a process improvement team**. The process operators are usually selected as a process improvement team. In some circumstances, it is wise to seek the opinion of a third party consultant, lending a degree of fairness and independent thinking to the judgments of the team.

4. **Document the process**. State the process in an orderly manner; we will show how a flow chart can be produced to help understand the process better. Add all necessary explanations to each activity in the process and describe the output of the process in its simplest terms.

5. **Collect the data from the process**. Measure the quality characteristics of the product. In many commercial processes, the quality characteristic is easily identified, although it might have several dimensions. In the past, many processes have failed miserably because the wrong characteristics were being measured. Knowing what the customer cares about means a lot when deciding what to measure from the process.

6. **Analyze the data**. Use the appropriate statistical techniques to analyze the data. The statistical software eZ-SPC includes a comprehensive collection of analytical techniques that can be easily employed with the click of a button. Because the computer only knows the numbers that are entered, it is important to understand the causes of existing problems in the process and how the results of the analysis affect our judgment of the problem.

7. **Develop a goal for improvement**. After understanding the current status of the target process, set a realistic goal for improvement. By thoroughly understanding the process, we can get to know its strengths and weaknesses. No matter how smoothly the process seems to be running, we will find ways for improvement. This is a continuous challenge for the team, because it seems to deny the conventional wisdom that

says "If it ain't broke, don't fix it". That cliché works for a few old-fashioned processes, but not for the majority of today's complex, multi-faceted systems that must undergo constant improvement and reliability growth.

8. **Take actions for process improvement**. Find the appropriate tools and techniques to make the process perform better. Once the idea of process improvement is instilled throughout the workforce, basic statistical tools like cause-and-effect analysis and Pareto analysis will become habitually used for process improvement.

9. **Evaluate results and compare them with the original objectives**. After taking actions to improve the process, goals often have to be rethought in light of newly obtained information from the analyses. In many applications, it may be necessary to repeat steps 6 and 7 until the goal is perfectly understood and fully achieved.

10. **Standardize the improvement actions**. Document the improvement activities for future reference, and standardize them for internal use.

11. **Write down the lessons learned**. Plan a meeting with the quality team and start the course of process improvement over again. Review all the activities and evaluate what lessons can be learned from previous actions. Change worker training methods if necessary to improve performance.

From a technical point of view, statistical process control can be seen as a toolbox, and achieving continuous process improvement requires knowing what tools are available in the toolbox, knowing their primary functions, and knowing how to select the right tool for the right task. The successful process manager will know the basics for process control: the concept of quality, customer, process, and technology, the concept of statistical thinking along with statistical tools and techniques, and the quality management system. The greater understanding of quality improvement leads to optimal implementation of statistical process control for reduction of variability in the process and the outputs.

A successful modern company must use effective methods to increase efficiency, effectiveness, quality, and productivity for better competitiveness. Throughout the organization, continuous process improvement must be recognized as an essential strategy. This book emphasizes the use of statistical thinking and statistical techniques to monitor the process and improve the performance of the process for better quality.

In our daily activities, variation is all around us and variation is present in everything we do. Identifying, characterizing, quantifying, controlling, and reducing variation provides us with new opportunities for improvement. The statistical thinking behind quality improvement is based on the assumption that all work is interconnected and variation is present and measurable in every process. We will show in this book that the techniques of statistical analysis that are applied in process control are quite basic and can be easily learned.

## 1.8   THE GOAL OF STATISTICAL PROCESS CONTROL

The ultimate goal of statistical process control is to reduce the variability in the process. We want to detect the assignable causes of variability in the process as early as possible. A manufacturing process that fails to do this will risk mass producing defective products that no one wants to buy. We want to eliminate the assignable causes and take appropriate actions to make sure the process will stay in control. In brief, statistical process control is a means to reduce variability in the process, to make the process mean close to the target value, to minimize the total quality cost, and to encourage company-wide involvement in quality improvement.

You can't fix every quality problem in one easy swoop. Process control requires constant attention to what is going into and coming out of the process. Quality improvement comes bit by bit, and for big companies, this requires total involvement not just from quality experts, but from line workers, process foremen and all of the top managers. Quality is more of a journey than a destination.

### The Myth of Quality

It is not easy to define quality. The word means different things to different people, and might signal anything from reliability to performance to luxury. Among scientists and engineers, the most widely accepted definition for quality is "fitness for use" given by Joseph Juran in 1974. Here is how other notable experts in quality control have defined the term:

- "Quality means conformance to requirements." (Philip Crosby, 1979)

- "Uncontrolled variation is the enemy of quality." (Edward Deming, 1980)

- "Quality is the characteristics of a product or service that bear on its ability to satisfy stated or implied needs." (American Society for Quality, 1987)

- "Quality is the loss a product imposes on society after it is shipped." (Genichi Taguchi, 1980)

- "Quality means doing it right when no one is looking." (Henry Ford, 1930s)

- "Quality is not an act, it is a habit." (Aristotle, about 350 BCE)

## 1.9   THE EIGHT DIMENSIONS OF QUALITY FOR MANUFACTURING & SERVICE

Everything we've said so far refers to quality as a positive feature, but we can see the word can be construed in different ways. Garvin (1987) introduced eight quality dimensions for manufactured products: performance, features, reliability, conformance, durability, serviceability, aesthetics, and perceived quality. We will explain what each of these words mean in terms of improving quality and then we will illustrate how they pertain to different processes.

### Eight Quality Dimensions for Manufacturing

1. **Performance.** A product's performance is judged on how well it completes its primary operating functions.

2. **Features.** A product's features typically help or compliment its performance, but do not constitute its main function. Like performance, features have measurable, objective attributes and are usually easy to compare between competing products.

3. **Reliability.** The reliability of a product refers to the likelihood it will not fail in performing its tasks. Reliability might be measure in time or in usage, and is a vital characteristic for durable goods.

4. **Conformance.** Most products must have operating features and characteristics that conform to industry standards. If the product does not conform to an established standard, it might be ruled to be defective.

5. **Durability.** A product's durability might be measured in its lifetime interval of usage. Durability might simply mean the total amount of use you get from it before its performance fails or degrades.

6. **Serviceability.** When a product needs adjustment or repair, serviceability is the speed, consideration and efficiency received by the customer. For less durable products, this might refer to a product replacement policy.

7. **Aesthetics.** How the product looks and feels to the customer is typically a subjective quality referred as its aesthetic appeal. This is a matter of form over function, and consequently it is difficult to measure in many cases.

8. **Perceived quality.** Since a product's performance and durability can seldom be observed directly, the customer can make a judgment through

the product's perceived quality and reputation. For most products, perceived quality is guided by advertising, peer approval and word of mouth.

Services have more intangible elements than manufactured products in general. Services are produced and consumed at the same time, for example, a call center service. Services are heterogeneous by ranging from simple to complex, from customized to standardized, and so on. Services are impossible to store so they are unique, for example, a medical service. In addition to the eight quality dimensions for manufactured products, Garvin (1988) also introduced eight quality dimensions for services: time, timeliness, completeness, courtesy, consistency, accessibility, accuracy, and responsiveness.

## Eight Quality Dimensions for Service Industries

1. **Time.** Service time is judged primarily on how fast it is provided. Waiting time and processing time must be reduced to satisfy customers.

2. **Timeliness.** Service timeliness is determined by the timing of service. In the ER example, the timing with regard to Cardiopulmonary Resuscitation is a critical factor in saving a patient's life.

3. **Completeness.** Service completeness is important to reduce costs required to finish any incomplete services dispensed to customers who were compelled to return. Complimentary services and supports must be available to the customers to ensure completeness of service.

4. **Courtesy.** Courtesy refers to the attentiveness, geniality and respect given to customers by the service provider. Customers can be just as upset with bad manners as they are with incompetence.

5. **Consistency.** Consistency of services refers to keeping the same level of service in any situation. A pizza franchise is careful to provide a standard recipe and price guide for all franchise outlets because customers expect the pizza to taste the same and cost the same no matter which outlet they order it from. This is an important quality dimension to the chain-food restaurant industry.

6. **Accessibility.** Accessibility of services refers to how easily customers can access the service. Both time and convenience are critical to accessibility. For banks, this means providing ATM machines at shopping malls.

7. **Accuracy.** Accuracy of service refers to the accountability of the service provided, including the input of personal data in the system, correct billing, and so on. If you have ever had your name misspelled or your sale receipt mistyped by a service provider, you know how frustrating the lack of accuracy can be.

8. **Responsiveness.** Responsiveness of service refers to how proactively service providers respond to customer needs. This requires the industry to anticipate things that can change or go wrong with the service, and initiating system changes immediately after receiving complaints or criticisms.

Even though the categories seem distinct, it can be challenging to match them easily for any product or service. Depending on the kind of industry, some of these dimensions can be overlapping and interdependent. The eight dimensions of quality for products and services that are listed in this section might not fully characterize all the aspects of quality for every conceivable product or service, but taking them into consideration will provide us with a basis of understanding what quality means.

Next, we will try to apply the dimensions of quality to examples in both products and services, respectively.

## Example 1-9: Smartphone

To illustrate the different dimensions of quality in a product manufacturing industry, let's consider the quality of a smartphone. These devices combine capabilities of a cellular phone with PC-like functionality. Examples include the Apple iPhone, HTC Touch, LG CT810 Incite, RIM Blackberry and Nokia E72. The device category is not very clear, though. We will assume a smartphone is a phone that runs modern operating system software and displays a standardized interface for the user. A typical smartphone will have keyboard interface, Bluetooth capability, camera, e-mail, Wi-Fi (and web browsing) ability, MP3 playback and can download and run simple software applications.

**Performance**: For most customer's, the primary task of the smartphone is the built-in phone. We will consider everything else to be contained under "features". To complete this primary task, the phone must connect when we call someone, and we should hear the receiver's voice clearly. Also, the receiver should hear our voice clearly.

**Features**: The cell phone has added quality with every extra feature it adds beyond the phone capability. Of course, each feature has its own performance measure, so the quality dimensions can be nested within each other in this case. Product upgrades often focus on increasing the number of features the smartphone offers.

**Reliability**: A reliable smartphone works well 100% of the time. The battery does not fail during the lifetime of the device, and the key buttons always work perfectly. None of the features will fail if it is accidentally dropped on a hard surface.

**Conformance**: There are several requirements that the smartphone should conform to, including weight, size, digital or analog conformity, and most important, conformity with long-distance phone plans.

**Durability**: In the last ten years, we found that cell phones become outdated quickly and did not have to last longer than the one or two year cell phone plan offered by the dealer. Smartphones will probably have the same durability issues, but a durable case and battery are pertinent, especially to customers who use the product frequently.

**Serviceability**: If something goes awry with the phone, the customer should find a convenient way to replace the smartphone or have it repaired. Because these devices become outdated quickly, companies find replacement policies are more practical for almost all repair problems.

**Aesthetics**: Fashion now plays an important role in marketing smartphones, so designers spend resources to ensure the product looks good to customers. Along with color and shape, other qualities such as the button sizes, the screen interface and even the cell phone size and shape are now considered matters of aesthetics.

**Perceived quality**: Smartphone manufacturers work hard, especially through marketing, at establishing a good reputation for making high performing, reliable and attractive cell phones. Among the many reputable companies that make smartphones, Apple has established a high perceived quality in an important population niche.

• • •

## Example 1-8: General Hospital Emergency Center

Let's briefly return to the hospital emergency center to see how the eight dimensions of quality can be applied to a service process. Since the primary mission of the hospital is to save the patient, performance can be measured in terms of patient outcome and patient satisfaction. Consider how you would classify the services the hospital emergency room offers in terms of the different quality dimensions:

- The service time refers to how long it takes until the patient receives all of the proper medical services before being discharged.

- The service timeliness refers to how appropriate treatments were applied at an appropriate time, including cleaning, bandaging, anesthetizing, dosing and discharging.

- The completeness of service implies the patient was stitched up completely or set on track to finish treatment after a follow-up appointment (e.g., surgery) is made.

- The consistency of service requires medical personnel to give equal treatment across different shifts in the emergency room and at all stations leading up to treatment and discharge.

- The responsiveness of nurses, courtesy of receptionist, and respect shown by all medical people is important in judging the quality of any service, especially when the customer is expected to be extremely vulnerable in some situations.

- Medical services must be easy to access. Ease of parking, clarity of signs, and the process entry function of the emergency facility can all determine success or failure for the medical service.

- Registration billing, personal information entry, and medical services must be accurate. These are sometimes not considered crucial in judging the quality of the emergency service, but this part of the industry can greatly influence the success of the service provider.

Clearly, this service-oriented example does not fit in easily to the eight boxes provided to describe the different dimensions of quality. But these two examples should provide you with insight on how we match up the eight quality dimensions with an actual industry, whether it is a traditional manufacturing industry or a more incongruent service industry.

• • •

## Quality versus Grade

Sometimes, comparable products are sorted by grade, which is totally different from quality. Grade is a category assigned to products or services having the same functional use but different technical characteristics. The more and the better features are promised, the higher the Grade. With hotel chains, for example, Holiday Inn is lower grade than Hyatt Regency Hotel. But Holiday Inn does not necessarily have lower quality than Hyatt Regency Hotel. In the automobile industry, for another example, Hyundai Motors Company's Sonata GLS and SE have different grades of engine, transmission, suspension/chassis, exterior features, and interior features. They have different grades, but not necessarily different levels of quality.

## 1.10   THE COST OF (POOR) QUALITY

If the quality of product is low or the product is defective, additional cost will be required to rework or repair the product to meet the intended specifications. Excessive variability in the process will cause waste. Quality costs are associated with production, identification, prevention, or repair of the products which do not meet the requirements. We also need to reduce the quality costs while improving the quality of products and the process. The quality cost consists of four categories: prevention cost, appraisal cost, internal failure cost, and external failure cost.

**Prevention Cost** - This cost is associated with activities for preventing defective products through the product life cycle. In other words, prevention costs are associated with efforts to make the product right the first time. They include costs in new product review, product design, quality education and training, quality planning and quality data analysis.

**Appraisal Cost** - This cost is associated with activities for measurement, evaluation, inspection, and tests of incoming materials, products, and components. They include costs in inspection and test, audit and maintenance of equipment.

**Internal Failure Cost** - This cost is associated with activities for rework or repair prior to shipment of products. They include costs in scrap, downtime, retesting and reinspection, and yield losses. These costs will naturally disappear if there are no defects in the product.

**External Failure Cost** - This cost is associated with activities for handling the products which are claimed by customers in the market. They include costs in customer complaints, customer returns, warranty charges, product recalls and some less visible costs due to customer dissatisfaction. These costs will be negligible if there are no nonconforming products shipped out for sale.

Quality costs will vary between different industries. A semiconductor manufacturer might incur quality costs that are up to half of its sales income. A fast-food restaurant might incur a cost of less than 5% of its gross sales.

## 1.11   WHAT DID WE LEARN?

- Statistical process control entails monitoring, controlling and improving a multi-stage process through statistical analysis.

- Variability in the process output consists of assignable cause variation and chance-cause variation.

- A primary goal in statistical process control is to eliminate assignable cause variation.

- The first control chart was developed in 1924 by Walter Shewhart of Bell Labs and Western Electric.

- A process is defined to be stable according to fixed target values and control limits.

- Continuous process improvement can be implemented with an 11-step procedure.

- Quality can be described using eight basic properties: performance, features, reliability, conformance, durability, serviceability, aesthetics, and perceived quality.

- The dimensions of quality for product manufacturers can be perceived differently by service providers.

- The grade of a product is determined by the kind of features, not the quality of them.

- The costs of quality can be split into four categories: prevention cost, appraisal cost, internal failure cost and external failure cost.

## 1.12   TEST YOUR KNOWLEDGE

**1.** Process control techniques can be used to eliminate which of the following from a process

    a. target values
    b. natural variability
    c. common cause variation
    d. assignable cause variation

**2.** In 1924, Walter Shewhart constructed the first

   a. bootlegging operation

   b. control chart

   c. push-up bra

   d. experimental design

**3.** In building a house, which of the following does not qualify as a controllable factor?

   a. amount of water used to mix cement for foundation

   b. amount of days work is delayed by rain

   c. number of builders assigned for framing house

   d. grade of shingle used for roof

**4.** If you buy a product with all the bells and whistles, you are increasing what aspect of your purchase?

   a. grade

   b. quality

   c. reliability

   d. conformance

**5.** Which of the following represents a step in continuous process improvement

   a. Cutting the fat to save money

   b. Reducing salaries to save money

   c. Documenting process with a flow chart

   d. Naming the process

**6.** Who said "Uncontrolled variation is the enemy of quality"?

    a. Edward Deming

    b. Walter Shewhart

    c. Vince Lombardi

    d. Joseph Juran

**7.** Golf ball manufacturers must be vigilant to ensure their product meets required specifications of the United States Golf Association. This refers to which dimension of quality?

    a. Performance

    b. Reliability

    c. Conformance

    d. Durability

**8.** In Example 1-5, which of the following represents an external failure cost to the juice manufacturer?

    a. Cost of high quality fertilizer

    b. Cost of fencing to keep out pests

    c. Cost of water used on berries

    d. Cost of replacing defective bottles sold to customers

**9.** In Example 1-4, which of the following represents an prevention cost?

    a. Cost of training pharmacist to run process

    b. Cost of advertising pharmacy

    c. Cost of disposing of defective pills

    d. Cost of customer loss due to poor service

**10.** In Example 1-3, which of the following attributes of the car-rental company represents the quality dimension of aesthetics?

    a. The rented car will start correctly the first time.

    b. The rental car agents are dressed cleanly and neatly.

    c. The customer queue is always short

    d. There is always someone at the service desk ready to help.

# CHAPTER 2

# GRAPHICAL DISPLAY OF DATA

The goal of this book is to help you learn statistical tools for quality improvement without having to learn statistical theory or mathematical proofs. Every statistical process decision is based on evidence, or data gathered from a process. For the variety of problems you will come across, *data* will be in the form of numbers that describe the output of a process. Typical data will include measurements used to verify whether the process output meets the required specifications of the product dimensions or performance. Measurements collected will tell us whether a process is in control, or in other words, whether a process is producing a reliable, consistent output at a steady pace.

Quality improvement techniques can be effective only if our understanding of the collected data is sound. Deciding what to measure in a process can be a significant challenge in its own right, but in this chapter, we'll take for granted that we already have decided what process output best characterizes the product quality. We will further assume we know how this output is to be measured, so our intent here is to find out how this collected data can be summarized and presented for best effect.

In this chapter, we introduce you to the analysis of SPC data through computer graphics. We will do this with the help of the statistical software

*Basic Statistical Tools for Improving Quality.* By Chang W. Kang and Paul H. Kvam  **29**
Copyright © 2011 John Wiley & Sons, Inc.

called eZ SPC. This is a simple program in the form of a spread-sheet with commands selected from a drop-down menu, very much like Microsoft Excel. Unlike Excel, we will provide you with all the tools you need to analyze SPC data, and unlike Excel, it is provided to you for free.

As you learn new tools for SPC, along the way we will continue to introduce new eZ SPC commands. Data analysis is performed using graphical and numerical displays:

**Graphical Data Display:** No matter how sophisticated the computational SPC tools seem to be, a picture is still worth a thousand words. Data that are displayed graphically provide insight into the fundamental properties of a statistical process. By browsing over graphical displays as simple as a bar graph or a pie chart, you will learn about more specialized displays such as box-plots and, in a later chapter, specialized SPC-charts.

**Numerical Data Analysis:** Summary statistics give us an idea of the average data value and how the data tends spread out around this average. Using more sophisticated tools, we can decide whether or not the process being scrutinized is under control, and if it is not in control, numerical data analysis can help us decide how out of control it is, and what we should do about it.

In this chapter, we will go over the graphical display of data. Each chart or graph introduced here can be generated using simple menu commands of the eZ SPC software. This leaves numerical data analysis for Chapter 3, where we will introduce you to simple analytical tools for summarizing data, including computing and interpreting statistics.

## 2.1   INTRODUCTION TO EZ SPC

The eZ SPC software is developed by author Chang W. Kang and his research team at Hanyang University, Ansan, South Korea. eZ SPC which accompanies this book presents a spreadsheet-based program with menu-driven commands to allow you to perform basic data analysis, graphical analysis and other methods used in statistical process control. You can download eZ SPC for free at

$$http://www.hanyang.ac.kr/english/ezspc.html$$

Along with the program, you should download the Applications Folder, which contains all of the data sets introduced throughout the book. Once you run the downloaded file, eZ SPC will appear on your desktop, as well as in your Program Files folder. To start your eZ SPC session, just double-click the new eZ SPC icon on your computer desktop.

Once you open the eZ SPC, you will recognize the menu-based structure because it is similar to that used by Excel  and other popular spreadsheet programs. Tabs on the top line separate the program commands into their basic function groups:

**File Commands** (save file, open file, close file, print file, etc.)

**Edit Commands** (copy, paste, delete, insert, find, replace, etc.)

**Tools** (calculator, histogram, box plot, probabilities, etc.)

**Graph Commands** (16 different charts and graphs for SPC)

**Analysis Tools** (Statistical methods for one or more samples)

To illustrate the computational tools you can wield in data analysis, we will start by opening the file LCDP.ezs in your applications folder. This file contains data from the manufacturing processes of two companies. Using this example, we will walk you through the basics capabilities of eZ SPC. You can open the file from the File command tab, but if you open the file by double clicking on the LCDP.ezs icon, make sure you choose to open the contents of the file with the eZ SPC.

---

**Example 2-1: LCD Panel Quality Data**

In the opened file, the two columns of data represent an important quality measurement of the distance between marks of a tape carrier package (TCP) bonded to liquefied crystal display (LCD) panels that were manufactured by two companies (A and B). The data are measured in millimeters (mm). The TCP is a packaging method of wireless bonding to the printer circuit board (PCB), which is applied to the Large Scale Integrated (LSI) Circuit. In this case, the specification limits are fixed at

$$25.22 \pm 0.01 \text{ mm},$$

and any LCD panel that has a distance measure outside the interval (25.21 mm, 25.23 mm) is considered defective. For each company, 50 panels were sampled, and the distance between recognition marks of the TCP was carefully measured. If the distance is outside the allowed specification limits, the data-signal delivery defect occurs because of miss-alignment at the bonding areas when the TCP is attached onto the LCD panel.

Once you have opened the LCDP file, left-click on the Graph menu tab (or, alternatively, hit ALT - g) to see your menu options for data analysis. Your display should look like the Graph menu in Figure 2.1.

Next, we will use eZ SPC to summarize the LCDP data set graphically. We want to understand if both companies are manufacturing LCD panels

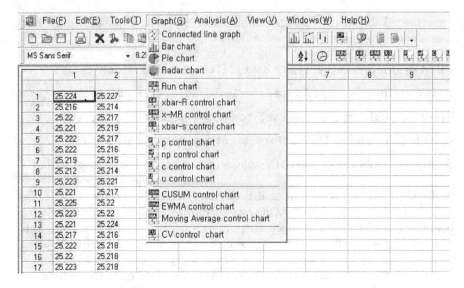

**Figure 2.1**    eZ SPC Graph menu display for LCD data.

that meet the required quality specification. In addition, we would like to compare the quality standards between the two companies. In Chapter 4, we will revisit this example and use analytical methods to answer questions about the evident quality differences in the two companies.

• • •

## 2.2   QUALITATIVE AND QUANTITATIVE DATA

Under the Graph menu tab, we see 16 different options of graphical display. Don't worry; we are only going to look at a few of them for right now. Some tools are for qualitative data (bar chart, pie chart and radar chart) and others, under the Tools menu tab are meant for quantitative data. These two data types are distinguished next.

The kind of data we run across in SPC problems can be categorized as qualitative and quantitative. To a statistician, qualitative data analysis is a less informative way to describe a set of data. Qualitative essentially implies the data cannot be explained and compared as ordered numbers. Qualitative data must rely on categories that uniquely and sensibly parse the information. For example, we might have an outcome of color emission, categorized into groups "yellow", "red" and "blue". We can't arbitrarily assign numbers to code the color groups, and then compare those numbers. Why not? If we assign 1 to yellow, 2 to red and 3 to blue, we have unintentionally implied that the difference between red and yellow is half of that between blue and yellow. That could lead to some embarrassing mistakes.

For qualitative data, one of the variables in the graph has no implied order, which lends a measure of vagueness to the picture. A bar graph, for example, will display the qualitative data on the horizontal axis (or X-axis), the order being alphabetic, and information on the vertical axis (or Y-axis) will generally be quantitative.

*Quantitative data* usually come in the form of informative measurements instead of simple attributes. Compared with qualitative data, quantitative (or *numerical*) data allows us to employ more powerful graphical tools and find out more about the collected information. With a numerical variable, order is important and differences in the ordered value can have spatial significance. The distance measure between the TCP recognition marks for the LCDs represents an example of numerical data. In this example, the distance measurement of 25.23 mm has a significant correspondence to the distance measurement of 25.21 mm, because both measurements are at the specification limit.

The graphical procedures for quantitative data are featured under the Tools menu tab. We will initially focus on basic methods to graphically summarize quantitative data: bar graph, pie chart, Pareto chart, radar chart, histogram, box plot, and scatter plot. These graphical features allow you to discern how the selected data are distributed according average value, the diffuseness of the data and where observations seem to cluster.

## 2.3   BAR CHART

With data categorized into a fixed number of groups, a bar graph displays the frequency of observations corresponding to each category of some variable. To implement the bar graph, the data should be entered in two columns: one for the category headings on the horizontal axis, and the other for the category frequency on the vertical axis.

### Example 2-2: Returned Restaurant Orders

Below are data collected by a restaurant in order to identify reasons for returned orders and use this data to implement process improvements. In a month-long study, the restaurant tabulated 101 returned food orders, and categorized the reason for returning the food into 5 major categories, with two return orders that did not fit in any of the five given categories. The data are found in the file **Restaurant.ezs** in your applications folder.

| Reason for Return | Frequency |
| --- | --- |
| Wrong order | 41 |
| Tainted or blemished | 36 |
| Arrived Late | 9 |
| Poor taste or quality | 7 |
| Cooked incorrectly | 6 |
| Other | 2 |

In this case, the variable is *Reason for Food Return*. From the Graph menu tab, the Bar Chart menu tool produces the following simple chart to visually compare the frequency of different reasons the food order was returned indicated on the vertical axis. The output is printed in Figure 2.2. The bar chart is the most basic graphical display in the toolbox, and can be used to compare the frequency of categories of the data, to recognize patterns or trends, and to get a quick, intuitive idea about the data without doing anything complicated.

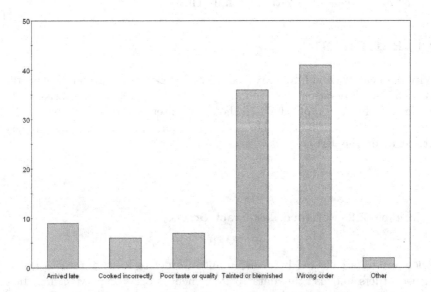

**Figure 2.2**  Bar chart displaying frequency of the various reasons for food order returns at a restaurant.

## 2.4   PIE CHART

Similar in function to the bar chart, the *pie chart* is a circular graph to show frequencies, but now the plot features the *relative frequencies* of the observations with regard to the other categories. The pie chart of the restaurant data from Example 2.2 is illustrated in Figure 2.3. The pie chart is useful to compare the proportions of the data that belong to each category, instead of the actual amount. This will be preferred over the bar chart if the proportions are more informative than the actual frequency count. In this case, the frequency count is 101 food returns, which probably has significance in the study, but the pie chart still helps communicate how frequent each reason for return is compared to the other given reasons.

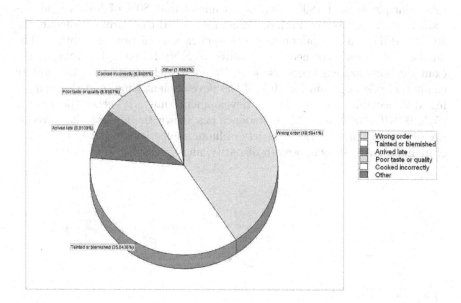

**Figure 2.3**   Pie chart displaying relative frequency of reasons for food order returns at a restaurant.

## 2.5   PARETO CHART

A *Pareto chart* is based on a bar chart with the categories plotted in descending order. Now the horizontal axis variable is re-arranged so there is an implied order in both axes. The Pareto chart is especially useful to show problems in a prioritized order, or to distinguish a *vital few* from the *trivial many*. Along with the charted bars, the diagram also includes a line that shows the cumulative frequency of the observations (from left to right). Note: unlike the previous procedures, the Pareto chart is listed under the Tools menu bar.

In process control, numerous sources of defects, error and product variability might be identified in a manufacturing process. This problem occurs in a large variety of applications, including biology, economics, sports and business. It is not uncommon that among dozens of identifiable problems, only a small subset of them seem to cause the most trouble. These are the "vital few" problems among the "trivial many", and this is the basis to the 80 - 20 rule: that only 20% of the identified problems cause 80% of the damage to the process.

The Pareto chart is an excellent tool to identify this kind of enigma in a process. The Pareto chart is a relatively new tool in SPC, but it is an old standby in economics. The Italian economist Vilfredo Pareto first thought of this principle around 1896 when he estimated that 80% of Italy's land was owned by 20% of its population. Sometimes called the Pareto Principle, the 80 - 20 Rule is a key factor in many applications of process control. The numbers 80 - 20 are not necessarily suited to every Pareto-type problem. For example, software engineers use a 90-10 rule which posits that the 90% of computer code accounts for 10% of the development time, and the remaining 10% accounts for 90% of the development time. Whether the ratio is 95/5, 90/10, 80/20 or 75/25, experience has shown these particular percentage differences characterize a variety of human experiences, including studies of manufacturing defects, wealth disparity, and several social statistics.

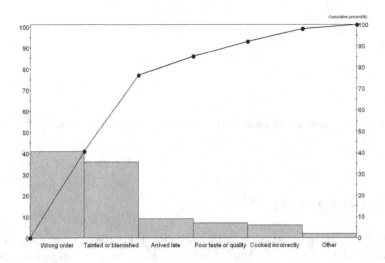

**Figure 2.4**   Pareto chart displaying ordered frequency of the various reasons for food order returns at a restaurant.

For the data from Example 2.2, the Pareto chart is especially useful. Out of the 101 returned food orders, the chart in Figure 2.4 shows explicitly how two causes account for over 3/4 of the problems. From the Pareto analysis, the restaurant manager might consider spending more work on creating a food order system in which the wait staff's written orders cannot be misinterpreted by the cooking staff. It will be more challenging to address the problem caused by food orders that were returned because the food was tainted or blemished. This might be a minor improvement made by the cooking staff, or it might indicate a more troubling problem with kitchen cleanliness or food quality.

In Figure 2.5, a different restaurant that serves fast food uses the same Pareto principle to diagnose problems associated with returned food orders. Over the course of one month, 71 of the reorders were broken into five different causes. The frequency of occurrence for the five potential problems are more evenly distributed, which refutes the Pareto principle. This result doesn't excuse the fast food restaurant in regard to the returned orders logged in the data. It only implies that the categories used in explaining the reasons food orders are returned failed to flesh out a dominant cause. For process improvement, the process manager should consider alternative ways of describing the problems associated with returned food orders.

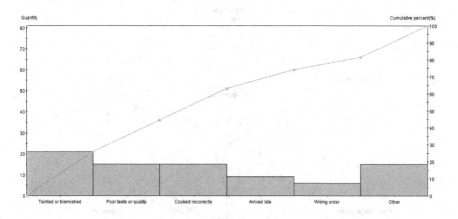

**Figure 2.5**    A new Pareto chart displaying frequency of the various reasons for food order returns at a fast-food restaurant.

There are other problems that might indicate that the Pareto principle is not valid for the data. If the "Other" group is one of the tallest bars in the chart, the category should be broken down into two or more new groups, if at all possible. In our example, we displayed 101 observations that were categorized into six categories. The number of observations should be over 50, as a rule-of-thumb. If you have only three categories, it might be hard to illustrate the Pareto effect when it exists in the data. If you have ten or more categories, it is probable that most of them will have few if any frequency counts, so the chart will still illustrate the Pareto principle, despite the potential confusion brought on by the excessive number of categories.

The Pareto chart will indicate dominant causes for defects or problems in a process output, but it will not distinguish these potential problems according how much they will cost to fix. For our restaurant example, if reducing the number of tainted or blemished orders is a more straightforward problem that will be much less expensive to solve when compared to solving the problems of incorrect orders, there is good reason to approach that problem first, despite the fact that it is not the primary category in the Pareto chart.

## 2.6   RADAR CHART

A radar chart, also known as a spider chart, is another type of visual display of categorical data with several axes. Each axis represents one of the categories, which protrude outward from the center to signify the frequency of each category. The radar chart is useful for comparing two or more processes using the same categories. The position of the axes is arbitrary, so the angles of the radar chart are not really informative.

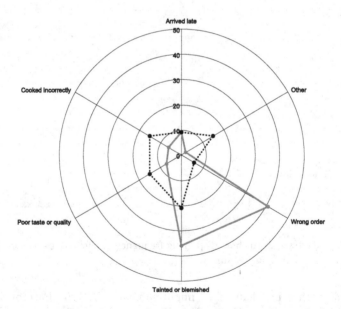

**Figure 2.6**   Radar chart comparing ordered frequency of the various reasons for food order returns at a regular restaurant (black line) and fast food restaurant (gray dashed line).

In Figure 2.6, the radar chart displays the returned-food order frequency data for Example 2.2 summarized in `Restaurant.ezs` (solid black line) along with the fast-food restaurant illustrated in Figure 2.5 (dotted gray line).The radar chart illustrates how the problem of returned food orders differs for the two restaurants. The Pareto principle is suggested in the black line by the spike and asymmetry of the radar chart. The chart for the fast-food data, on the other hand, generates a more compact and convex shape because the reasons for returned orders are more equally spread out among the possible categories. In this case, the two sets of data do not have equal sample size; it will be easier to interpret the radar chart if all of the sample sizes are the same.

## 2.7  HISTOGRAM

A histogram is a graphical display of frequency data with bars. With a histogram we can check the distribution of data. To illustrate how the histogram works in eZ SPC, we return to the LCDP data in Example 2.1. In your LCDP spreadsheet, select the column of data pertaining to Company A, and then select the Histogram tool from the Tools menu tab. It should be shaped like Figure 2.8. From the picture of the histogram, we can infer the distribution of distance measurements in the range of 25.2118 mm and 25.2262 mm; these are actually the midpoints of the first and last of the ten classes. The class centered at 25.2198 mm appears higher than the others; the class with the greatest frequency is typically called the histogram's *mode*.

The histogram also shows how the data are dispersed across the *range* of the data (the numbers between the smallest and largest value), and whether the data are *symmetric* (distribution looks similar on right and left side) or *skewed* (categories appear to decrease or increase on one side of the distribution, making the histogram asymmetric). Figure 2.7 shows a simple example of how histograms can illustrate symmetry and skew.

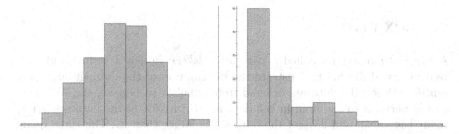

**Figure 2.7**    Histogram of a symmetric set of data (on the left) and a right-skewed set of data (on the right).

The histogram for the LCDP data appears in Figure 2.8. In this histogram, three observations are slightly outlying to the left, suggesting they have unusually small distance measurements compared to the general sample. The vertical axis reveals the frequency in each category - 13 observations in the modal class, for example.

The histogram might be our best tool to turn a long column of numbers into an informative visual display. With less than 10 numbers in that column, the histogram plot gives very little helpful information, but if the set of the data has over 30 observations, the plot is surprisingly good at informing us about how the data are distributed. While it does not present us with sharp numbers that tell us about the central tendency or variability of the data, the picture communicates these ideas and more. In some examples, we can spot a perfect bell-shaped curve in the histogram. This makes statisticians happy, at least relative to other emotions they display in public. In other examples, we might spot a skew, or maybe an outlier represented by a bump somewhere on the horizontal axis that is separated from most of the other observation groups.

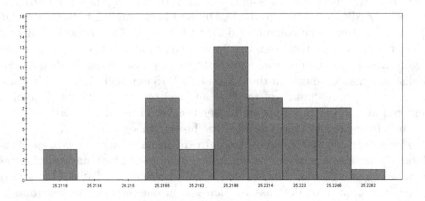

**Figure 2.8**   Histogram of LCDP distance measurement data for Company A.

## 2.8   BOX PLOT

A *box plot* (sometimes called a *box-and-whiskers plot*) is a simple and convenient visual display to illustrate the distribution of the data through *percentiles*. When the data are ordered from smallest to biggest, the 75th percentile pertains to the item in the data for which 75% of the data set is less than that value, and 25% of the data set is larger than it. The three quartiles (25th, 50th, 75th percentiles) are meant to split up the data into four equal sized parts, with a quarter of the data set in each. The box plot shows five critical features of the data, including three percentile values. Figure 2.9 shows a box plot - a segmented box with lines sticking out the bottom and top - and how to interpret it. The five critical features of the data are:

1. The smallest observation in the data

**2.** The largest observation in the data

**3.** The upper quartile of the data, which is the 75th percentile.

**4.** The lower quartile of the data, which is the 25th percentile.

**5.** The median, or middle-ranked value of the data (50th percentile).

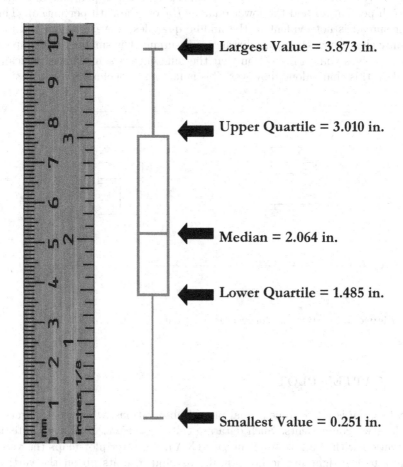

**Figure 2.9**  Five summary statistics reported by the box plot.

Box plots are also helpful when you want to compare two or more samples. From Example 2.1, we can compare the distance measurements of Company A and Company B in LCDP.ezs by constructing two adjoining box plots and comparing their five key features.  To make the box plots for the LCDP example, open the file LCDP.ezs and select the data from both the distance measurements in columns A and B, then left-click on the Tools menu tab and select

Box Plot.

You should produce a box plot similar to that shown in Figure 2.10. Note that the procedure produces a plot along with a summary sheet, so you don't have to guess the box plot statistics from the plot itself. From the box plot, you can notice that the distance measurements from Company B are spread out more (the sample range is greater), especially in the upper quartile (above the 75th percentile) and the lower quartile (below the 25th percentile). This larger spread is not evident in the middle quartiles, however. The box plot not only shows how the mean distance measurement is smaller for Company B, but it gives you an indication that the difference is somewhat significant. Based on this chart alone, however, this is not very conclusive.

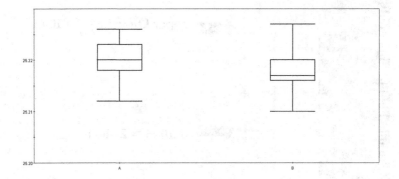

**Figure 2.10**   Box plot for measurement data from Company A and B.

## 2.9   SCATTER PLOT

A *scatter plot* is a two-dimensional plot of observations, where each observation has two components. Such data are called *bivariate*. If we designate an observation with the 2-component pair (X,Y), a scatter plot maps the value X units to the right in the horizontal axis, and Y units up on the vertical axis. To illustrate the scatter plot, we return to the LCDP data in Example 2.1. The temperature data for the LCDP data are contained in the file `Temps.ezs`. The distance measurements for each company are taken in 50 consecutive time periods, so the order of the data in the spreadsheet reflects the order in which the measurements took place. Figure 2.11 shows a scatter plot of the distance measures for Company A along the vertical axis, and the sample order, listed 1 to 50, is listed on the horizontal axis. As you glance at the figure, the plotted points don't seem to have any distinct, recognizable pattern. This shows that the distance measure from a sample at any point in time is not conspicuously affected by the sample that came before it or after it. In other words, the sequential observations don't seem to be related, or correlated to each other in terms of sample order. A *correlation* between the X and Y variable implies that the placement of one (say X) will influence the outcome of the other (Y). A statistic used to measure correlation will be discussed in Chapter 3.

Without the scatter plot, it is difficult to display bivariate data, compared to regular single-component data. For example, if each measurement from the data has two variables, we could make a histogram for each of the variables to summarize the overall distribution. But this marginal summary of the bivariate observations ignores the potential relationship (or correlation) the two measurements might have with each other.

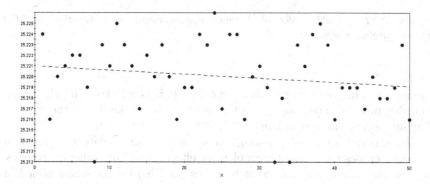

**Figure 2.11**    Scatter plot of Company A distance measurements as a function of time of manufacturer.

Another example is illustrated in Figure 2.12. In this example, there is a explanatory variable that is measured along with the distance measurement: temperature of the printer circuit board after assembly. In this case, temperatures range from 30 to 50 degrees Celsius, and the scatter plot in Figure 2.12 indicate that the quality decreases if the PCB temperature is high. This is an example of statistical correlation that will be further investigated in Chapter 3.

## 2.10  CAUSE AND EFFECT DIAGRAM

A *cause and effect diagram* (or *Ishikawa diagram*) is an illustration that is used to explore potential or real causes of quality problem. Causes are arranged in four or five major inputs, such as material, machine, people, method, and environment. From their relationships and hierarchy of sub-causes, the cause and effect diagram can help search for root causes and identify areas where the problem may be related. The cause and effect diagram is one of the most

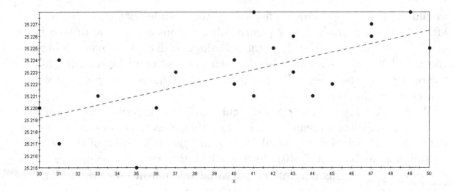

**Figure 2.12**   Scatter plot of distance measurements as a function of PCB temperature after assembly.

applied tools in the problem solving process. It is also linked with the brainstorming process. The cause and effect diagram is also known as the *fishbone diagram* because its resemblance to the skeletal frame of a fish.

The fishbone shape of the diagram helps to communicate how different potential causes for a specific problem or effect come from different sources. In a brainstorming session, extra branches (or ribs) can be easily added or deleted, leading to a highly dynamic problem solving session. To make a cause-and-effect diagram, we first state the problem in the form of a question. If the problem solvers study the reason why the problem exists, team members can focus on different root causes that can answer the question in different ways. The following example shows a real problem that a manufacturer faced with unexpected stock outs that can delay their manufacturing process.

## Example 2.3: Unexpected Wire Stock Outs

The Nordson Corporation manufactures systems to apply adhesives in industrial settings. A large portion of the work performed at their manufacturing plant in Norcross, Georgia is the production of hoses allowing glutinous substances to flow through and be applied. Wiring is required to control temperature and application of the adhesives, and hence, wire is a large component of the production process. When wire is received from shipping and dispatched to various ongoing production processes, it tends to get lost or not be fully accounted for in the plant. Wire often arrives in large spools, commonly containing 1,000 feet or more. However, most projects require only about 40 feet of wire to complete.

Unfortunately, workers tend to discard spools that hold wire less than this amount and it becomes necessary to splice wire from a new spool for projects that require small amounts of wire. Furthermore, discarded wire is not entered into Nordson's supply chain database and often is just simply lost. Such acts are common in the Norcross plant and cause unexpected stock-outs of wire parts, which can halt the production process until new wire can be ordered and received to resume the pending project. Other residual causes of wire stock-outs include the fact that wire arrives in large spools and is not pre-spliced. As a result, wire is inevitably lost because ongoing projects often need varying amounts of wire and the common multiples of wire used can vary widely at times. The Cause and Effect Diagram for the event of unexpected wire stock-out for hose manufacturer is displayed in Figure 2.13.

Before specific causes are hastily put on the diagram, make sure the input categories are logical and clearly stated. If there are too many categories, some groups should be coupled. If there are only two categories, you will end up with a pathetic looking fishbone, so make sure there are three or more input categories.

· · ·

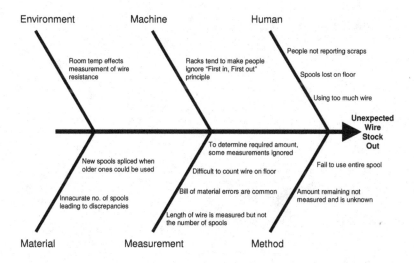

**Figure 2.13**    Cause and Effect Diagram for the event of unexpected wire stock-out for hose manufacturer.

## 2.11   WHAT DID WE LEARN?

- Statistical process control relies on both graphical and numerical data procedures.

- The included software, called eZ SPC, will provide all of the graphical displays of data and statistical analysis procedures needed for this book.

- Qualitative data cannot be ranked like quantitative (numerical) data can. Basic plots used in statistical process control include the bar graph, pie chart, Pareto chart, radar chart, histogram, box plot and scatter plot.

- Cause and effect diagrams are used to explore potential quality problems in a process.

- The 80-20 rule states that 20% of a process's problems cause 80% of the damage.

- The box plot includes five critical summary values from the data: the smallest and largest values along with three quartile values.

## 2.12    TEST YOUR KNOWLEDGE

**1.** Which of the following procedures are graphical tools for summarizing and displaying data?

a. Pie Chart

b. Phono Graph

c. Standard Deviation

d. Box Plot

**2.** In horse racing, it has been shown that the top 20% of the jockeys win 80% of the races. This is an example of

a. Unfairness in sports

b. Skewness

c. The Pareto principle

d. A frequency histogram

**3.** Which of the following statistics are reported in a box plot?

a. The sample mean

b. The sample median

c. The sample variance

d. The sample range

**4.** How is the histogram related to the bar chart?

a. They are both charting relative frequency.

b. The histogram is a special case of a bar chart.

c. They both have an ordered horizontal axis.

d. The bar chart is a special case of a histogram.

**5.** After constructing an eZ SPC figure, right-clicking on the figure will allow the user to

a. Change fonts on axes and figure labels.

b. Copy or save the figure as a jpg.

c. Change the colors of bars or lines in the figure.

d. All of the above.

**6.** How is a pie chart different from a bar chart?

   a. The categories in a pie chart are ordered according to size.
   b. The pie chart requires more colors (or shades of gray).
   c. The pie chart computes only relative frequency.
   d. The pie chart cannot be constructed using eZ SPC software.

**7.** To compare two or more processes in terms of categorized output, we can use

   a. A pie chart
   b. A histogram
   c. A radar chart
   d. A Pareto chart

**8.** Which of the following tools are offered under the eZ SPC Tools tab:

   a. Calculator
   b. Bar chart
   c. Radar chart
   d. Scatter plot

**9.** Which of these will cause problems in constructing an effective Pareto chart?

   a. Most of the defects are categorized in the category "Other".
   b. There are only six categories of defects defined.
   c. There are fewer than 30 total observations.
   d. One of eight categories of defects contains 80% of the observations.

**10.** If the increase in one input variable (X) causes the decrease of the output variable (Y), these variables are

   a. Independent
   b. Positively correlated
   c. Negatively correlated
   d. Explained graphically using a scatter plot.

# EXERCISES

**2.1**  Open the file `coffeeshops.ezs` to see the average monthly sales (in units of $10,000) for 100 coffee shops across the United States. Create a histogram for the data and comment about the distribution of the monthly sales.

**2.2**  An airline company wants to investigate the reasons for delayed take-offs related to their commercial jets. Data are collected for a specific month, and over 600 events were recorded in the table below. What is the best way to graphically display the results in order to find out what are the major reasons for take-off delay?

| Reason for Delay | Frequency |
| --- | --- |
| Maintenance | 150 |
| Ramp | 85 |
| Security | 120 |
| Weather | 54 |
| Boarding | 67 |
| Flight operations | 102 |
| Other | 60 |

**2.3**  The 2009 monthly rainfall totals for Atlanta, Georgia are given in the file `Atlanta.ezs`. Open it and generate a scatter plot of the 2009 data by assigning the X-axis to the first column (month labels) and the second column to the rain amount (in inches). Next, compare the 2009 rainfall data with the average monthly amounts using the `Connected Line Graph` under the Graph menu. How did the year 2009 compare to the average?

**2.4**  Use the same data file `Atlanta.ezs` from the previous exercise, and look at the fourth column, which contains the 2009 monthly electric bill (in dollars) for a small house in Atlanta, Georgia. Are the electric bill and monthly rainfall for 2009 at all related? Construct a scatter plot of the data using the 2009 rainfall for the X-axis and electric bill for the Y-axis.

**2.5**  From the 2010-2011 best hospital rankings by the *US News & World Report*, the top 5 best hospitals were Johns Hopkins, Mayo Clinic, Massachusetts General, Cleveland Clinic, and the UCLA Medical Center. For each specialty, the scores are given below. ENT stands for "Ear,Nose,& Throat" specialists. To compare the five hospitals in specialty areas, try a radar chart. Why is the radar chart appropriate?

| Hospital | Cancer | ENT | Gynecology | Urology | Heart |
|---|---|---|---|---|---|
| Johns Hopkins | 75.1 | 100 | 100 | 100 | 70.1 |
| Mayo Clinic | 79.1 | 78.4 | 99.6 | 87.1 | 88.1 |
| Massachusetts General | 57.1 | 80.5 | 82.1 | 50.6 | 65.3 |
| Cleveland Clinic | 52.6 | 68.3 | 88.7 | 99.1 | 100 |
| UCLA Medical Center | 51.8 | 61.9 | 65.3 | 79 | 54.9 |

**2.6**    Find an example of the 80-20 rule that you have experienced at work or at home. You might be surprised just how frequently the Pareto principal permeates our daily lives. In his book, *The 80/20 Principle: The Secret to Success by Achieving More with Less*, Richard Koch finds examples in stock investments, crime statistics, tax revenues and energy allocation (to name just a few). Tyler Perry connected the 80-20 rule to marriage success in his 2007 movie *Why Did I Get Married*. As a personal example, we have found that less than 20% of the students in our classes will utilize over 80% of the available office hours.

Try to find an example that can be quantified. Gather enough data so you can graphically display the results in a Pareto Chart. What does the chart communicate to you? Does this example represent a problem that needs to be fixed? Or does it just represent a part of your life in which 80 percent of what you do may not count for much?

**2.7**    The Group of Twenty (G-20) Finance Ministers and Central Bank Governors was established in 1999 to bring together systemically important industrialized and developing economies to discuss key issues in the global economy. The inaugural meeting of the G-20 took place in Berlin, on December 15-16, 1999, hosted by German and Canadian finance ministers. Population growth rate is defined as the average annual percent change in the population, resulting from a surplus (or deficit) of births over deaths and the balance of migrants entering and leaving a country. The growth rate is a factor in determining how great a burden would be imposed on a country by the changing needs of its people for infrastructure (e.g., schools, hospitals, housing, roads), resources (e.g., food, water, electricity), and jobs. Open the file `population.ezs` to find the population growth for the 19 countries listed in the report, and use a bar graph to compare rates between countries. *Source: CIA World Factbook - Unless otherwise noted, information in this page is accurate as of February 19, 2010.*

# CHAPTER 3

# SUMMARIZING DATA

In this chapter, we highlight some basic statistical terms and techniques that will be used in process control, and we will skip statistical theory that is not necessary to sufficiently perform these analytical tasks. We will do this, in part, with the help of the accompanying statistical software, eZ SPC. As we learn techniques for statistical process control, we will continually learn about more helpful methods of analysis based on numerical data. In this section we will only consider the most basic properties of the data, and leave the advanced analytical techniques for later chapters.

The two most fundamental properties of a data set are its *central tendency* (location) and its *variability* (scale). You might not know exactly what we have in mind when you hear these words, and unfortunately, there is no uniformly adopted measure of either quality. For example, in measuring the central tendency of a group of numbers, the "average" value of the data, the "middle" value of the data and the "most typical" value of the data might each point to a different answer. We will be quite specific in what we need to understand about central tendency and variability in this chapter. First, let's see how considerations of location and scale are applied to a set of data. Suppose we have a set of measurements for 10 products. The measurements

*Basic Statistical Tools for Improving Quality.* By Chang W. Kang and Paul H. Kvam **51**
Copyright © 2011 John Wiley & Sons, Inc.

are the number of nonconformities that were found on each product through visual inspection:

$$0 \quad 8 \quad 5 \quad 3 \quad 4 \quad 0 \quad 7 \quad 1 \quad 0 \quad 14.$$

There is no unique middle value because the ordered sample has an even number:

$$0 \quad 0 \quad 0 \quad 1 \quad 3 \quad 4 \quad 5 \quad 7 \quad 8 \quad 14.$$

However, the two middle values (3, 4) are probably more representative of what we are thinking when we consider a typical value from this sample.

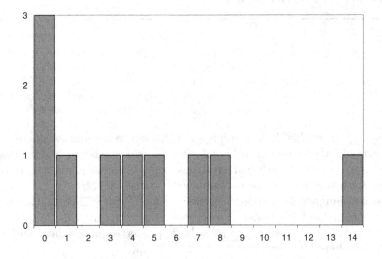

**Figure 3.1**  Histogram of sample data.

Figure 3.1 displays the simple histogram of the ten observations from this sample. The histogram represents the graphical display of the frequency of the data, showing the number 0 occurs three times, while the other nine observations {1,3,4,5,7,8,14} occur only one time. The gaps in the histogram give us a picture of where the center of the data occurs and how the ten observations are spread out around that center. But overall, the graph cannot give us a dependable measurement for the location of the middle or the scale of variability that we need to accurately summarize a data set.

To define our measures of location and scale, we need to represent each measurement from the sample in a general but convenient way. We will denote the first observation from the sample as $x_1$, the second as $x_2$, and so on. The nonconformity data can be rewritten as

$$x_1 = 0, x_2 = 8, x_3 = 5, x_4 = 3, x_5 = 4,$$

$$x_6 = 0, x_7 = 7, x_8 = 1, x_9 = 0, x_{10} = 14.$$

With this general expression for the sample values, we can more easily describe the numerical measures of central tendency and variability.

## 3.1   CENTRAL TENDENCY

To convey the essence of a set of data to your audience, the first idea to get across is the central tendency. There are two popular ways of measuring the central tendency of the data: the sample mean and the sample median. It's important to know that they do not always agree, and when they don't, it is good to know why.

The sample mean (or average) is nothing but the sum of the measurements divided by the number in the set. We will denote the average by putting a bar over the letter that represents the data:

$$\bar{x} = \frac{x_1 + x_2 + \ldots + x_n}{n} = \frac{1}{n} \sum_{i=1}^{n} x_i.$$

In science and engineering, the "$\Sigma$" sign denotes the addition operator, but we will try to avoid these kinds of math symbols wherever possible. For the nonconformity data listed above, the sample mean is computed simply as

$$\bar{x} = \frac{0 + 8 + 5 + 3 + 4 + 0 + 7 + 1 + 0 + 14}{10} = 4.2.$$

In this example, the sample mean (4.2) is a number that doesn't exist in the data. This is often the case with the sample mean, and might be a bit confusing to people who are not good with numbers. Specifically, we're thinking about the kind of people who read that the average number of children per family for a particular region is 2.3 and wonder what 0.3 children look like. Some other measures of central tendency, such as the media, tend not to have this problem as much.

Similar to the sample mean, the sample median is defined as the middle value of the ordered data. That is, half the data are smaller, and half the data are larger. For example, if the data consists of 99 observations, the median would rank as the 50th smallest out of 99, with 49 observations being less than the median and 49 observations being larger. If the sample consists of an even number of observations, then there is no unique middle-ranked value, so we typically take the average of the middle two values. That was the case for the nonconformity data, where 3 and 4 are the middle two values (the 5th and 6th smallest out of 10), so the sample median is the average of 3 and 4, or 3.5. Notice that with an even number of observations, the median can be a number that doesn't exist in the data.

The mean and median both measure a central tendency, but different kinds of central tendency. With the mean, every observation has equal representation in the data set, so the presence of a single extreme value in the data might have an undesired influence on the mean. For example, suppose we have a different set of data based on $n = 5$ values:

$$x_1 = 12, x_2 = 32, x_3 = 16, x_4 = 25, x_5 = 20.$$

The sample mean is computed as $\bar{x} = 21$, which seems to be a fair representative of the data, even though the number 21 never actually appears in the original set. Suppose we add another single observation ($x_6 = 205$) to the data set. This observation is unlike the five others, and it changes the sample mean to $\bar{x} = 51.67$, which is substantially larger than every observation in the set except that one. We would label $x_6$ as an extreme value or an *outlier*.

The median value of the data set does not have this sensitivity to extreme values like the mean does. This is because the data are not equally represented. Instead, one number from the set is a chosen representative - the middle one. If we use the same data set, (12, 32, 16, 25, 20), the median is 20. Like the mean, this seems to be fairly representative, and the number 20 actually occurred once in the data set. If we add a single number ($x_6 = 205$) to the data set again, the median is 22.5, the average of the two middle-ranked values ($x_5 = 20$, $x_4 = 25$). In this case, the median has hardly changed after adding the outlier to the data set.

The advantage using the median as a measure of central tendency is that the median is not sensitive to a single extreme value. One disadvantage the median, compared to the mean, is the difficulty in computing it. Rather than adding up a list of numbers, the median requires that the data be ordered, which can be relatively time consuming to the computer. For modern computers, this is rarely an issue, even for large data sets.

## 3.2   VARIABILITY

After the central tendency, the variability of a data set is the next most important characteristic to display. To measure how the data is spread out around its center, there are different measures that reflect different aspects about how the data is dispersed. The most primitive is the *sample range*, which is the distance between the biggest number in the sample and the smallest. For any two equal sized data sets, the data sampled from a more dispersed population should have a bigger range, in general. So the range statistic is used for detecting variability in a process.

But the sample range has some shortcomings. For one thing, your statistic will be determined by the extremes in the data, ignoring the spread within the set. For example, with the previous set of data based on $n = 5$ values

$$x_1 = 12, x_2 = 32, x_3 = 16, x_4 = 25, x_5 = 20.$$

The sample range is computed as the smallest observation (12) subtracted from the largest observation (32) which is 20. If we add the extra observation ($x_6 = 205$) to the data set, the sample range changes to 205 - 12 = 193. The extremeness of $x_6$ has an enormous affect on the range. Besides the sensitivity to outliers, the sample range naturally increases as sample size increases, so large samples tend to have larger range values, whether or not they are spread out. Comparing the variability of two samples with the sample range would only be fair if the sample sizes are equal.

## Variability in Terms of the Standard Deviation

The standard tool used in business and science for measuring *sample variability* is the sample variance statistic ($s^2$) and the square root of the sample variance ($s$), called the sample *standard deviation*. The sample variance is computed by first finding the sample mean $\bar{x}$ and then going back to the data set computing

$$s^2 = \tfrac{1}{n-1} \sum_{i=1}^{n} (x_i - \bar{x})^2.$$

This book doesn't show off a lot of math formulas, so when a formula is featured, you can consider it a very important equation. First, let's distinguish the sample variance and the sample standard deviation. Statisticians use the sample variance for mathematical reasons, but it becomes difficult to explain because the units for $s^2$ are actually the squared value of the units of the data. For example, if the quality characteristic of a service process is the number of minutes required to finish serving the customer, then the unit of measurement is minutes, but the sample variance is measured in minutes-squared (or minutes$^2$, in math terms). The standard deviation, on the other hand, is measured in the original units of the sample.

The sample standard deviation, computed through $s^2$ as

$$s = \sqrt{\tfrac{1}{n-1} \sum_{i=1}^{n} (x_i - \bar{x})^2},$$

is difficult to compute, compared to the sample range, but it is superior in terms of measuring how disperse the data are in the sample. The only time we rely on the sample range is when the data set is small (less than 10) because the sample variance needs a minimal number of observations before it becomes a good estimator of dispersion.

To illustrate how the sample standard deviation is computed, we will return to the artificial data set of $n = 5$ values:

$$x_1 = 12, \; x_2 = 32, \; x_3 = 16, \; x_4 = 25, \; x_5 = 20.$$

Recall the sample mean is computed as $\bar{x} = 21$, so to compute the sample variance (on the way to getting $s$), we need to take the differences $x_i - \bar{x}$, square them to make $(x_i - \bar{x})^2$, then add them up and divide by $n$ - 1 = 4:

$$\frac{(12 - 21)^2 + (32 - 21)^2 + (16 - 21)^2 + (25 - 21)^2 + (20 - 21)^2}{4},$$

which equals 61. The standard deviation is $\sqrt{61} = 7.81$.

Businessmen and scientists have learned to use standard deviation as a rule of thumb. For some common types of data, it turns out that around 2/3 of the data will be within one standard deviation of the sample mean. That is, if we have a set of data and compute the sample mean ($\bar{x}$) and the sample standard deviation ($s$), then the rule of thumb tells us that approximately 68% of the observations are likely to be in the interval ($\bar{x} - s, \bar{x} + s$). We can go further. It turns out that in many data sets, we can expect around 95% of the data to be within two standard deviations of the sample mean. If we go out further, nearly all of the data, over 99%, will be within three standard deviations of the sample mean (see Figure 3.2).

The theoretical (and typically unknown) standard deviation of the population being sampled is denoted by the Greek letter $\sigma$. The famous management innovation "six sigma" is related to this and will be discussed more in future chapters.

## Variability in Terms of the Interquartile Range

Although the standard deviation is the most commonly used measure of dispersion in a sample, there are other measurement options aside from the range of the data. The interquartile range (IQR) is defined to be the distance between the 75th percentile in the data and the 25th percentile in the data. Recall that the quartiles (25th, 50th, 75th percentiles) break the data set into four equal sized parts. The observations in the interquartile range represent half the data set, with the other half of the data split equally on both sides. The IQR is the range of the "middle half" of the data, so it is not sensitive to extreme values like the sample range would be.

If we revisit the figure of the box plot in Figure 2.10 from the last chapter, the quartiles are plainly marked and the IQR is just the length of the box

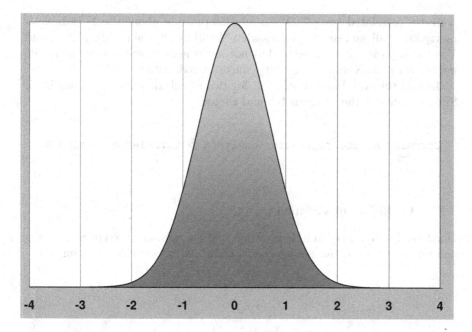

**Figure 3.2**    Normal distribution with vertical lines representing the length of one standard deviation.

(ignoring the protruding lines). In our example, we can see the IQR for Company A's measurements is larger than that of Company B. You can verify this by looking at the data sheet that accompanies the box plot. The summary data includes the regular box plot statistics as well as the IQR.

## Variability and Correlation

To understand the idea of *correlation*, let's go back to the scatter plot example in Chapter 2. The scatter plot allows us to see how bivariate data (data with two components instead of one) are distributed. When an observation is represented by a pair of numbers (X,Y), the correlation provides an analytical measure of how closely they relate to each other. If Y goes up and down as X does, then the pair is positively correlated. If Y seems to disobey X by going down when X goes up, the two variables are negatively correlated. The correlation statistic $(r)$ is standardized between $-1$ and $1$. If $r = 0$, then X and Y seem to take no notice of each other, and are considered uncorrelated.

In Figure 3.3, the three scatter plots represent examples where the correlation between the X variable on the horizontal axis and the Y variable on the vertical axis differ. The correlation is positive $(r > 0)$ because Y seems to go up, in general, when X goes up. In the bottom plot, X and Y are nearly perfectly correlated ($r = 0.99$), so any information about the X variable is pretty much all we need to predict what Y will be. In the middle plot, X and Y are still positively correlated, but not as strongly. Here, $r = 0.57$. In the uppermost plot, X and Y are nearly uncorrelated, with $r = 0.06$.

To find the correlation coefficient for two matched columns of data in eZ SPC, look under the Analysis tab and choose

```
Correlation and regression analysis ▷ Correlation analysis.
```

### 3.2.1    Coefficient of Variation

One final description of sample variation that is commonly used in engineering and finance is called the sample *coefficient of variation*, which is defined as

$$CV = \frac{s}{\bar{x}}.$$

CV is simply the ratio of the process standard deviation over the process mean. This measure is useful when dealing with experiments or processes in which the mean is positive and changes from one setting to the next. Obviously, if stays the same for two processes, the CV can be no more informative than the standard deviation in comparing the process outputs.

In manufacturing, when a product is improved to increase yield, it is not uncommon for the data to spread out more as the mean increases. Comparing a large manufacturing process (with high yield) to a small manufacturing process (with modest yield) might lead the analyst to think the smaller process is more stable because the variance of the small yield is so much less than the variance of the high yield of the larger process. In general, this is not a fair comparison if we base it only on the standard deviation or variance. The coefficient of variation evens out the effect of the standard deviation by describing it as a portion of the mean value. In finance, the coefficient of variation is used to compare the risk (related to standard deviation $s$) versus the average returns ($\bar{x}$) on two different investments. A lower the ratio of standard deviation to mean return means a better tradeoff of risk-return.

## 3.3    STATISTICAL DISTRIBUTIONS

In the last two sections, we looked at how to summarize a data set in terms of its sample mean and standard deviation. Different data sets can be easily distinguished if we know these two critical statistics. These numbers don't

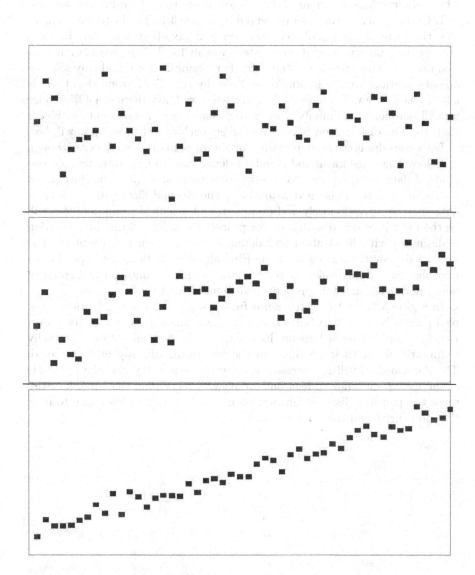

**Figure 3.3**    Three examples of correlated data where the correlation coefficient $r = 0.03$ (top), 0.44 (middle), and 0.97 (bottom).

tell us entirely how the data are distributed, but they give us a good idea. In this context, *distribution* describes how the data are spread out. It shows where observations are more likely and less likely to be found in the sample.

Let's look back at the data presented in Figure 3.1. The histogram tells us that three out of the ten observations are zero, six observations are between one and 10, and no observations are larger than 15. This represents information about the distribution of the data. For example, if we randomly selected one observation from this sample, we know by this distribution that there is a 30% chance we will sample a zero. Likewise, we know there is a 40% chance we will sample an odd number, and a 10% chance we will get a number bigger than 10. The distribution tells us how likely each one of these picks will be.

Let's consider how two populations might differ in distribution even though they have the same mean and standard deviation. To illustrate, we pick two kinds of data distributions that are well known in statistics: the Normal (or Gaussian) and the Poisson distribution. The Normal distribution describes data that are spread closely and evenly around a central mean, and this will be the most common distribution for process data. The Normal distribution is discussed with illustration and detail in Section 3 of this chapter. The Poisson distribution, along with the Binomial distribution, are important in treating count data (called attribute data) such as counting the number of nonconformities, and this property is demonstrated later.

In Figure 3.4, the left histogram is from a sample of 100 observations sampled from a Normal distribution, and the right one is a sample of 100 observations sampled from a Poisson distribution. The Normal data are generally symmetric about their middle, and the histogram will appear bell-shaped. The Poisson data will not necessarily have this symmetry, and the right side of the histogram tends to lean out (or skew) farther than the left side. Still, these two pictured distributions were constructed for this illustration to have the same mean and the same variance.

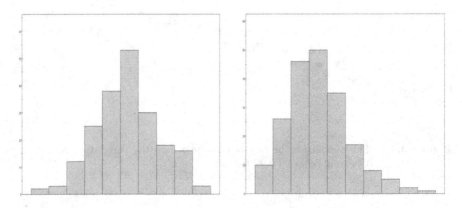

**Figure 3.4**   Histogram of (a) Normal and (b) Poisson distributed data.

## Normal Distribution

We won't be taking an expedition to learn about all of the different distributions that statisticians use in probability and data analysis, but the Normal distribution is special. When you heard about the *Bell Curve* or that your test scores in college were graded "on a curve", this usually referred to the curve of the Normal distribution. So why is the Normal special? Why does that last sentence seem like an oxymoron?

The reason has to do with something called the *Central Limit Theorem*. What the Central Limit Theorem states is that if you have a properly sampled set of data, the histogram (or distribution) of the sample mean ($\bar{x}$) will look more like the Normal distribution as the sample size gets bigger and bigger, even if the distribution of the underlying data have nothing to do with the Normal distribution. This works not only on typical measurements (time, area, distance, weight) but even on attribute data, for which every observation can be a binary response (like zero/one, yes/no, fail/succeed). Attribute data is as non-Normal as you can get.

To show how the Central Limit Theorem works, suppose we have attribute data, and from a sample of $N$ products, let

$$Y = \text{the number out of N of the products}$$
$$\text{that are found to be defective.}$$

If we are inspecting a product lot in which 20% of the items are defective, then each product we inspect should have the same 1-in-5 chance of being found defective (we'll pretend the lot will regenerate itself after sampling and is essentially infinite in size). As long as the sampler picks out products in a fair and *independent* manner, then this distribution for $Y$ is called *Binomial*. Here, independence between two sampled products means that the outcome of the first sampled product does not influence the sampling of the second one. Each item sampled item has a 20% chance of being defective, independent on what happened with the previous sample or the samples that follow.

To estimate the proportion of the population that is defective, we would use the proportion of defective items in the sample:

$$\frac{Y}{N} = \text{sample proportion.}$$

By using a one to represent a defective item and a zero to represent a non-defective item, Y is considered a sum of N zeros and ones. In a way, Y has the same qualities of $\bar{x}$, which is also a weighted sum of the data. In other words, we can treat Y from a sample of $N$ as we would treat a sample mean from a sample of the same size. Figure 3.5 shows how the distribution of the number of defects changes as the number of sampled products $N$ increases. The figure has four histograms, each based on 100 repeated samples (so the total number of inspected products in each is $100 \times N$) and each product has a 20% chance of being defective. We can see that as the sample size ($N$) increases, the histogram starts to look more bell-shaped. This is a ramification of the Central Limit Theorem.

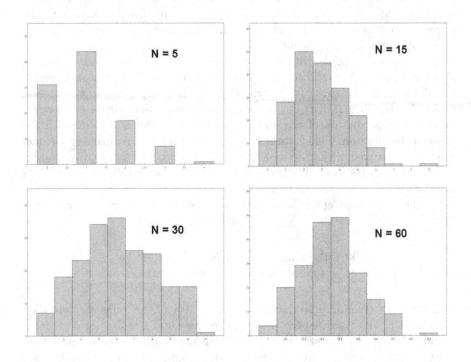

**Figure 3.5**   Histograms of Binomial data with $N = 5, 15, 30, 60$.

At this point, you should understand that sampled data can be treated as if they were randomly chosen from some particular distribution which we explore and discover through the histogram. In the case we inspect 30 items from a lot where 20% of the items are defective, our intuition tells us that we might expect 20% of our sample (six items) to be defective. But we also know there is randomness, so we won't get exactly six, at least not every time. If we inspect 60 items, we will expect to get twelve defects, but one should not find it so surprising to get anywhere from nine to fifteen; in fact, 74% of the time we will observe between 9 and 15 defects in this case. The Central Limit Theorem says that in the latter case, where we expect 60 items at a time, the histogram (centered around the expected value, 12) will look more bell-shaped than the histogram based on repeated trials where we inspect 30 items at a time. In the latter case, the histogram is centered around six (the expected number of defects out of 30), but it is slightly less bell-shaped.

## 3.4    DISTRIBUTIONS IN EZ SPC

Computing frequency probabilities for distributions can be cumbersome. Below, we present some simple examples of how we can use eZ SPC to make probability calculations for various problems and distributions.

### Example 1-8: General Hospital Emergency Center

Suppose the general hospital emergency center claims an 80% survival rate of heart attack patients if patients arrive hospital within 3 hours after detecting the first symptoms of a heart attack. In the previous month, it was determined that 12 heart attack patients arrived at the emergency center within this 3 hour interval. What is the probability that 10 or less patients survive?

To compute the answer, we don't have to learn all of the rules of probability that are needed to answer it. With the help of a computer, we only need to recognize the distribution of the data presented in the problem. In this example, each patient represents an independent trial, each having the same survival probability. In this sense, if $Y$ = number of patients (out of 12) who survive, then $Y$ has the binomial distribution. To compute any binomial probability, we can use eZ SPC. Left click on the Tools tab and select

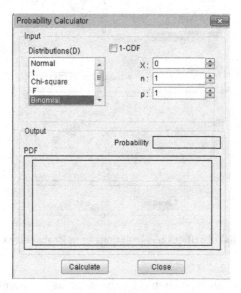

**Figure 3.6**   Probability Calculator Window in eZ SPC.

## Distributions

from the menu tab. The window that pops up is called the `Probability Calculator`. From this window (see Figure 3.6), select

## Binomial

from the left column that is labelled `Distributions(D)` and do the following:

1. Input 10 for $X$ (this tells eZ SPC to compute the probability of selecting 10 or fewer patients out of the total lot).

2. Input 12 for $n$ (so $n$ represents the total number of patients that were sampled, thus $X$ must never be larger than $n$).

3. Input 0.8 for $p$ (so $p$ represents the proportion of patients who survive, and $p$ must be selected between 0 and 1).

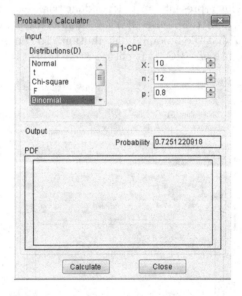

**Figure 3.7**  Binomial distribution calculations from Probability Calculator Window using $X = 10$.

The numbers you entered should be the same as the numbers that appear in the right-hand column of the `Probability Calculator` window in Figure 3.7. Click on the `Calculate` tab that appears on the bottom of the window, and the output probability should appear as in the Figure. With eZ SPC, we calculate the probability that up to 10 patients survive (out of 12) is 0.7251220918, or about 72.5%.

We know that there is a 72.5% chance that between zero and 10 patients will survive if their heart attack symptoms are detected within three hours of their arrival to the emergency center. What if we want to know the probability that *exactly* ten patients out of 12 survive? The eZ SPC `Probability Calculator` gives us only the *accumulated* probabilities for all possible frequencies less than or equal to ten. To get the probability of exactly ten, we have to be clever. Use the `Probability Calculator` again to find the probability that 9 or fewer patients will survive (repeat the procedure above in eZ SPC, replacing the input $X=10$ with $X=9$).

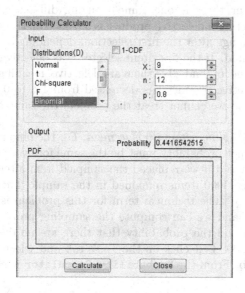

**Figure 3.8**    Binomial distribution calculations from Probability Calculator Window using $X = 9$.

To find the probability that exactly 10 patients survive, we can use the calculated probability that at most 10 patients survive (0.725) and *subtract* the probability that at most 9 patients survive (0.442) that we get from Figure 3.8. So the probability that exactly 10 patients survive is 0.725 - 0.442 = 0.283 (or 28.3%).

•  •  •

## Example 3-1: Injection Molding

A company produces small injection molding products. A lot includes 3,000 products and 1% of products (30 out of 3000) are known to be defective. The 100% inspection is employed if 1 or more defective products are found in the inspection of 100 products. What is the probability that 100% inspection can be waived?

We are counting defective products out of a lot in which the probability of detecting one is 1 in 100. This problem is similar to the binomial distribution example above, but there is one game-changing difference. Although the probability of selecting a defective product is 30/3000 or 1/100, this detection probability will change after the first selection, depending on whether or not a defective was found. If the first item selected was defective, then there are 2999 items left in the lot, and 29 of them are defective, meaning the probability of selecting a defective product has decreased from 1% to 0.97%. Although this change is small, we cannot treat the problem the same as we did in the previous example.

This is called *sampling without replacement*. For the binomial distribution, the defective product probability must be the same for every selection, which could only be achieved if we replaced the sampled item after every selection (thus ensuring that 3000 items remained in the sample, knowing that 30 of them are defective). The technical term for this problem is called *hypergeometric sampling*, and we can compute the sampling probabilities using eZ SPC. We need to find the probability that there are no defective products in this sample of 100. Left click on the Tools tab and select `Distributions` from the menu tab. From the `Probability Calculator` window, select

`hypergeometric`

from the left column that is labelled `Distributions(D)` and do the following:

1. Input 0 for $X$ (this tells eZ SPC to compute the probability of selecting 0 or fewer defective products out of the total lot).

2. Input 3000 for $N$ (so $N$ is the total lot size of the products, including both defectives and non-defectives).

3. Input 0.01 for $p$ (so $p$ represents the proportion of products in the lot of 300 that are defective).

4. Input 100 for $n$ (so $n$ represents the number we selecting out of the total lot of $N = 3000$)

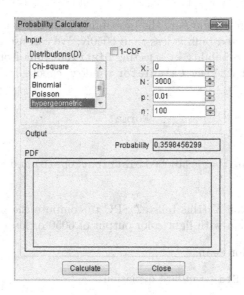

**Figure 3.9**    Binomial distribution calculations from Probability Calculator Window using $X = 10$.

The probability of getting no defectives in the a sample of 10 (out of 3000) is 0.3598456, or about 36%.

• • •

## Example 3-2: Fluorescent Light Bulbs

According to the U.S. Department of Energy, an ENERGY STAR qualified compact fluorescent light bulb (CFL) will save about 30 dollars over its lifetime and pay for itself in about 6 months. It uses 75 percent less energy and lasts about 10 times longer than an incandescent bulb. Light intensity is measured on the Kelvin scale (K). For the best reading light, the intensity should be near a natural daylight color, which is between 5000K and 6500K.

The Manufacturer produces CFLs with with an average output of 6000K. There is a slight variability between bulbs, however, and the distribution of output follows a normal distribution (with a bell-shaped curve) that has a standard deviation of 30K. Recall from earlier in this chapter that if the output light has a normal distribution, approximately 68% of the bulbs manufactured will be within one standard deviation of the mean, or in this case, they will be between 5970K and 6030K.

Suppose we want to compute the probability that the light color output of any randomly selected bulb is less than 6050K. We can use the Tools tab eZ SPC to find this probability by selecting `Distributions` from the menu tab, and from the `Probability Calculator` window, choose

`Normal`

from the left column that is labelled `Distributions(D)` and follow these three steps:

1. Input 6050 for $X$ (this tells eZ SPC to compute the probability of selecting an item with light color output of 6050 or less).

2. Input 6000 for `Mean`.

3. Input 30 for `Std` (standard deviation).

After clicking on `Calculate`, Figure 3.10 shows the output probability to be 0.952209, or about 95%. For the normal, eZ SPC also plots the bell-shaped curve that represents the distribution, along with the vertical line for the chosen $X$ value of 6050. In this case, 0.9522 represents the area under the curve to the left of that vertical line.

• • •

### Example 1-3: Renting a Car

At the airport car rental agency, it is determined that on average, five customers arrive per hour (during business hours). To help with assigning personnel and resources to the rental desk, the agency would like to know the probability that no more than once customer arrives to the rental desk between 10 AM and 11 AM on a given day.

By knowing five customers arrive per hour *on average* does not mean exactly five people must arrive between 10 and 11 AM. The number varies according to some distribution. For counting problems like this one, it is often assumed that the *Poisson Distribution* is used to model how the arrival frequency might change. Figure 3.4 showed a histogram of the Poisson distribution and contrasted it with the normal distribution. Using the same sequence in eZ SPC

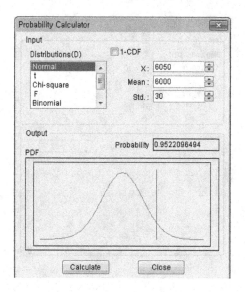

**Figure 3.10**    Normal distribution calculations from Probability Calculator Window.

Tools ▷ Distributions ▷ Poisson,

input $X=1$ and Mean=5, then click on Calculate. The probability is shown in Figure 3.11 as 0.040437682, or about 4%. It is highly unlikely the rental agency will see fewer than two customers per hour during business hours.

• • •

## Example 3-3: Gas Mileage

Residents of a metropolitan city are concerned with fuel consumption of their cars as the gas price increases. A survey study team collected mileage per gallon (MPG) data of intermediate size car (1800cc) in four metropolitan cities such as Los Angeles (LA), New York (NY), Seoul, and Tokyo. The data are presented in the file MPG.ezs in the Applications Folder. The team wants to compare MPG in four different cities. How can you compare them?

Each city data set has 50 observations. First, find the sample mean and sample standard deviation of the four sets of data and compare them directly. To do this, select

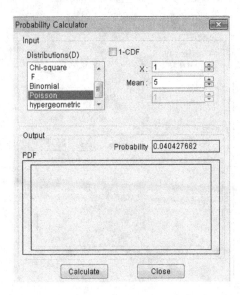

**Figure 3.11**    Poisson distribution calculations from Probability Calculator Window.

## Estimation for 1 - Sample Mean

from the Analysis menu tab. When a new window for **Select Sigma** appears, select the option for **Estimate**. This option is needed because the population standard deviation (denoted with the Greek letter $\sigma$) is unknown and must be estimated by the data using the sample standard deviation, $s$. Which city reported the best MPG ratings?

Next, construct histogram to show the distribution of data for each city and compare these graphical displays to your statistical summary above. It is difficult to compare four histograms at one glance. eZ SPC provides a feature **Tile vertically** under the Windows menu tab. For the best result, you can follow these steps:

1. Produce the histogram of MPG data from LA.

2. Edit the chart header title by using the right button of mouse.

3. Enter the scale of the x-axis manually: **max** = 50 and **min** = 0.

4. Repeat steps (1) – (3) for the other three cities: NY, Seoul, Tokyo.

5. Remove worksheets so that there are only four histograms open in eZ SPC.

6. Tile vertically (as described above).

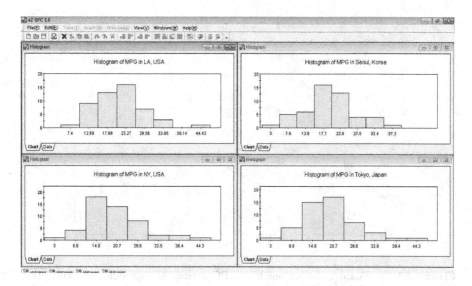

**Figure 3.12**   Histograms for automobile MPG ratings at four different cities: Los Angeles, New York, Seoul, Tokyo.

As a result, you should obtain the display in Figure 3.12. Another way to compare the means and variances graphically is to use the boxplot. If you select all four columns of data and then choose Box-plot under the Tools menu tab, you should get a display with four box-plots like the one featured in Figure 3.13.

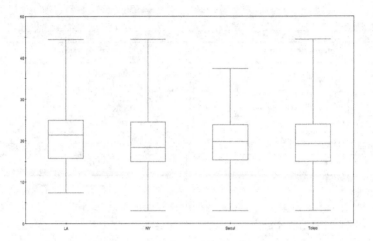

**Figure 3.13**    Boxplots for automobile MPG ratings at four different cities: Los Angeles, New York, Seoul, Tokyo.

## 3.5    WHAT DID WE LEARN?

- Two primary characteristics of a data set are its central tendency and variability.

- There are different ways of estimating unknown sample variability, including the sample range, standard deviation and interquartile range.

- The correlation coefficient measures how closely two variables relate to each other.

- The normal distribution describes populations that produce frequency histograms that feature a symmetric, bell-shaped mound.

- The central limit theorem guarantees that sample averages will have an approximately normal distribution if the sample size is large enough.

- If we sample products from an assembly line with the same probability of selecting a defective product in each turn, the number of defective products will have a binomial distribution.

- If we sample products from a finite lot that has a fixed number of defective products, the probability of selecting a defective product changes from turn to turn, depending on whether or not a defective was selected previously. This is the genesis of the hypergeometric distribution.

- The binomial distribution is based on sampling with replacement. The hypergeometric distribution is based on sampling without replacement.

## 3.6   TEST YOUR KNOWLEDGE

**1.** In which of the following lists of data is the sample mean equal to the sample median?

   a. −4, −1, 0, 1, 4
   b. 4, 8, 15, 16, 23, 42
   c. 1, 7, 8, 12, 16, 16
   d. 3, 13, 14, 15, 24

**2.** In which of the following sets of data is the sample variance equal to the sample range?

   a. 1, 1, 3.5, 3.5, 6, 6
   b. 0, 0, 0, 8, 8, 8
   c. 0, 1, 2, 3, 4
   d. 0, 4, 4, 4, 8

**3.** In an experiment, we roll 10 6-sided dice and write down the average number we observe on the 10 faces. We repeat this experiment 100 times and construct a histogram of the data, which turns out to look bell-shaped. This is due to

   a. The Pareto Principle
   b. The Central Limit Theorem
   c. The hypergeometric distribution
   d. The histogram

**4.** Open LCDP.ezs and use eZ SPC to find the sample mean of the distance between recognition marks of the TCP for Company B (second column).

   a. 25.21792
   b. 25.21726
   c. 25.21741
   d. 25.21815

**5.** Use eZ SPC to find the Interquartile Range (IQR) for Company B's distance measurements. The IQR is listed on the data sheet that accompanies a box plot.

   a. 0.0001
   b. 0.0040
   c. 0.0055
   d. 0.0140

**6.** Select the first 10 distance measurements for both companies (2 columns) and construct a scatter plot using eZ SPC. What is the correlation coefficient between these ten pairs of numbers?

    a. Less than zero

    b. Between zero and 0.30

    c. Between 0.30 and 0.60

    d. Larger than 0.60.

**7.** From Example 3-2, find the probability of selecting a light bulb that has light color output less than 5985K.

    a. 0.3085

    b. 0.5000

    c. 0.6477

    d. 0.6915.

**8.** From Example 3-2, find the probability of selecting a light bulb that has light color output more than 5990K.

    a. 0.3165

    b. 0.5000

    c. 0.6306

    d. 0.6477.

**9.** From Example 1-8, find the probability that at most 3 out of the next 6 patients arriving within three hours of detecting symptoms of a heart attack will survive. Recall that 80% of such patients will survive, in general.

    a. 0.0989

    b. 0.1818

    c. 0.7250

    d. 0.9011.

**10.** The Poisson distribution is typically used to model which of the following phenomenon?

    a. The girth of fish

    b. The Central Limit Theorem

    c. Arrival frequencies

    d. Sampling without replacement.

## EXERCISES

**3.1**  A pizzeria is interested in reducing the pizza delivery time to satisfy customers with better taste. The pizzeria collected the actual delivery time from shop to a customer for 50 orders. The data are expressed in units of minutes, and can be found in the eZ SPC file `pizzeria.ezs`. Find the average, median, range, variance, and standard deviation for the delivery times. How often is the delivery time within one standard deviation of the mean? That is, how many delivery times are longer than $\bar{x} - s$ but less than $\bar{x} + s$?

**3.2**  The number of patients arrived at a general hospital emergency department varies from day to day. The hospital collected data for the daily number of patients arriving for 40 days and the data are given in `ER.ezs`. Find the average, median, range, variance, and standard deviation. Look at the histogram of the data and decide if the distribution is *symmetric* or *skewed*.

**3.3**  According to the J.D.Power and Associates 2010 US Initial Quality Study (IQS), problems per 100 vehicles (PP100) are reported by the automobile company name. The scores are listed in `AutoReliability.ezs`. Find the average and compare the score for each automobile manufacturer with this average. Find 25th percentile, the 75th percentile and the interquartile range (IQR). Which company scores are above the 75th percentile?

**3.4**  A bank located in downtown Seoul is interested in serving more customers at each teller window. The service time by a bank teller approximately follows normal distribution with mean 93 seconds and standard deviation 40 seconds. The bank plans to reduce the service time to 60 seconds per customer. What is the probability that the service time is less than 60 seconds?

**3.5**  To describe the rate of arrival for customers at a bank, it is common to employ the Poisson distribution. Let us assume this works for the problem above, where customers arrive at a downtown bank in Seoul. If the mean arrival rate is one customer every 60 seconds (or 60 arrivals per hour), what is the probability that 50 or fewer customers arrive in the first hour?

**3.6**  A cell phone battery manufacturing company supplies ready-to-use batteries to cell phone manufacturing company for cell phone packaging process. Most batteries come ready to start but some require charging before packaging. This wastes time on the packaging process. The cell phone manufacturing company has the rule to reject the shipment of 1,000 batteries if one or more uncharged battery is present among 10 randomly selected batteries. If the probability of uncharged battery shipment is 0.2%, find the probability that the specific shipment is rejected.

**3.7**   Open `Atlanta.ezs`, which contains data we examined in the last chapter. The file contains both the 2009 monthly rainfall totals for Atlanta, Georgia (in inches) and the 2009 monthly electric bill (in dollars) for a small house in Atlanta. Instead of using a scatter plot, select `Correlation and regression analysis` ▷ `Correlation analysis` under the Analysis tab. What is the correlation coefficient for this data? Does this suggest there is a strong or weak relationship between the amount of rainfall and energy consumption for households in Atlanta, Georgia?

# CHAPTER 4

# ANALYZING DATA

In the last chapter, we learned different ways of summarizing data. The way a set of data is distributed can be a matter of scientific debate, and there are various distributions can be used in statistical process control. Two characteristics of the data, the mean and variance, are of primary importance. The sample mean provides us with our best estimate of a population's true average. If we only have five observations on which to estimate this mean, we would have less confidence in its accuracy compared to another sample comprising of 50 observations. With more observations comes more certainty.

Science is a systematic way of collecting data about nature and summarizing that knowledge into testable laws and theories. In this chapter, we investigate how to quantify this knowledge through the use of confidence intervals and tests of hypotheses. A scientist might devise a theory to make a claim about nature, but the claim remains valid only if it is backed up by empirical evidence. With evidence, more is better. Here are how some great thinkers have weighed in with regarding empirical testing in science:

- " When one admits that nothing is certain one must, I think, also admit that some things are much more nearly certain than others." – Bertrand Russell (1872 – 1970)

*Basic Statistical Tools for Improving Quality.* By Chang W. Kang and Paul H. Kvam  **77**
Copyright © 2011 John Wiley & Sons, Inc.

- "Experimental confirmation of a prediction is merely a measurement. An experiment disproving a prediction is a discovery." – Enrico Fermi (1901 – 1954)

- "...it doesn't matter how beautiful your theory is, it doesn't matter how smart you are – if it doesn't agree with experiment, it's wrong." – R.P. Feynman (1918 – 1988)

- "Errors using inadequate data are much less than those using no data at all." – Charles Babbage (1792 – 1871)

- "It is easy to lie with statistics, but it is easier to lie without them." – Charles Frederick Mosteller (1916 – 2006)

## 4.1   CONFIDENCE INTERVALS

We learned that the sample mean is what we use to guess or *estimate* the population mean. Even if we have no idea about how the data are distributed, the Central Limit Theorem implies that a histogram for $\bar{x}$ will be bell-shaped, and the distribution of $\bar{x}$ is approximately Normal. Once we compute the sample mean, how confident should we be that it is close to the unknown population mean? If we average 4,000 observations, I'm rather certain that my sample average is a good estimator for the population average. But if I only have 4 observations, I realize that, due to chance, the average of only four observations does not inspire the same confidence. As the sample size increases, we are gaining more confidence in the sample average being close to the population mean. This is because the average will vary less as the sample gets larger.

Specifically, the sample average ($\bar{x}$) based on $n$ observations can be shown to have a standard deviation that is $\sqrt{n}$ times smaller than the standard deviation of the individual sampled items. So the average from a sample of $n=100$ observations has a standard deviation $\sqrt{100} = 10$ times smaller than the standard deviation for the individual measurements. And less variability means more certainty. As an example, the probability that a sampled item from a Normal population has a 68% chance of being within one standard deviation of the population mean. But if we sample 9 observations, the sample mean has a standard deviation that is $\sqrt{9} = 3$ times smaller, and the chance of being in the same interval around the mean increases to 99.7%. We're more confident using $\bar{x}$ as an estimator for the population mean.

There are two ways statisticians gauge certainty in the sample average. A confidence interval is a pair of numbers representing a lower and upper bound for the population mean, based on the sample average. A test of hypothesis is decision tool, again based on $\bar{x}$, used to decide whether the population mean is too big, too small or just too unpredictable. In this section, we will discuss how confidence intervals can help quantify uncertainty in a simple way.

## Example 4-1: 2000 US Presidential Election

In the weeks upcoming to a presidential election, pollsters like Gallup and Zogby are using survey data to predict how the election voting will come out. For example, before the 2004 U.S.A. presidential election, Gallup's last poll predicted George W. Bush would achieve 51% of the vote and Zogby predicted 49%. The final tally showed Bush received about 50.7% of the vote, compared to John Kerry's 48.3%. News media and the public often criticize the pollsters for inaccuracies, and 2004 was an especially loud year by the critics. How can two professional polling groups come up with an estimate that disagrees by two percentage points? Does this show that the pollsters cannot be trusted?

Actually, there are important aspects of the polls that the media sometimes fail to report. For example, Gallup's poll predicted Bush would achieve 51% of the vote, but with a 3.1% margin of error. What does this *margin of error* mean? By adding this margin of error, Gallup was presenting a confidence interval for the predicted vote turnout.

In fact, it would be more correct to say Gallup predicted that Bush's tally would be in the interval

$$51\% \pm 3.1\% = (47.9\%, 54.1\%),$$

but even that doesn't tell the whole story. This interval also has a measure of certainty, which is left unsaid, but is typically assumed to be 95%. So the best way to interpret Gallup's result would be to say we are 95% certain that the percentage of voters who would vote for George Bush in the 2004 presidential election is between 47.9% and 54.1%. This is a 95% confidence interval, and its length is 54.1% - 47.9% = 6.2%. We can interpret this 95% certainty to imply that for every 20 poll predictions they make, on average one of those intervals is wrong. By using this correct way of viewing poll results, we can see that the Gallup and Zogby polls did not actually contradict each other, and the confidence intervals were not only informative, but this time they were entirely correct.

• • •

Confidence intervals represent an effective way of describing uncertainty in an estimate. There are two critical factors of a confidence interval: the length of the interval and the confidence level ascribed to it (usually between 90% and 99%, and frequently assumed to be 95%). Any statistical decision will have the caveat of uncertainty, otherwise it would be dogma and not science. To illustrate we will return to our LCD Panel data.

## Example 2-1: LCD Panel Quality Data

We return to LCDP.ezs, where the two columns of data represent the distance (in mm) between marks of TCP bonded to LCD panels that were manufactured by two companies (A and B). In this case, the specification limits are fixed at $25.22 \pm 0.01$ $mm$, and any LCD panel that has a distance measure outside the interval (25.21 mm, 25.23 mm) is considered defective. For each company, 50 panels were sampled, and if the distance is outside the allowed specification limits, the data-signal delivery defect occurs because of missalignment at the bonding areas when the TCP is attached onto the LCD panel.

To see if Company A is complying with the specs, we will analyze the 50 specimen measurements in the first column. To do this, select

Estimation for 1 - Sample Mean

from the Analysis menu tab. When a new window for Select Sigma appears, select the option for Estimate because the population standard deviation (denoted with the Greek letter $\sigma$) is unknown and must be estimated by the data using the sample standard deviation, $s$. You should have produced a new window with the following summary statistics:

| Confidence level | CI for the mean |
|---|---|
| 90.0% | (25.21921516 , 25.22082484) |
| 95.0% | (25.21905525 , 25.22098475) |
| 99.0% | (25.21873399 , 25.22130601) |

| Descriptive Statistics | |
|---|---|
| N | 50 |
| Mean | 25.22002 |
| Standard Deviation | 0.00339562 |
| Variance | 0.00001153 |

The lower half of the output (Descriptive Statistics ) presents some basic statistics: the sample mean ($\bar{x}$), the sample standard deviation ($s$), and the sample variance ($s^2$) based on $N = 50$ observations. The top half of the output computes two-sided confidence intervals for the mean. The confidence intervals (CIs) have different lengths because they provide different degrees of confidence. A 90% interval allows that 1-in-10 intervals made in this manner will be wrong. By allowing a 10% error rate, this interval is narrower than a 99% confidence interval, which will be conservatively wider to ensure that only

1-in-100 intervals will be in error. The thing about statistics and about life, is that we will never know for sure if this was one of those 1-in-100 intervals.

The first thing to notice about the results is that the range of the specification limits lies entirely inside the 99% confidence interval. We are at least 99% certain the true mean distance between recognition marks for company A lies within the specified limits. If we repeat the estimation using the second column of data for Company B, the 99% confidence interval for the mean is (25.21661064, 25.21922936). This also lies completely within the specification limits.

• • •

Because confidence intervals are a powerful way of describing uncertainty in an experimental outcome, we can construct them across all sorts of applications and all sorts of data.

## Confidence Intervals for the Difference in Two Population Means

We can compare the results of Companies A and B using confidence intervals as well. In this case, select both columns of data and choose

### Estimation for 2 - Sample Mean

from the Analysis menu tab. This procedure provides a summary for both samples along with a confidence interval for the mean difference between the two populations. For example, the LCD panels from Company A have a slightly larger average for the distance measurement compared to those from Company B. The 90% confidence interval for the difference of means between Company A and B is (0.00096238, 0.00323762). An important detail we can garner from this interval is that zero is not in it. That is, if you conjectured the mean distance measurements are the same for both companies, this interval shows with 90% certainty that this is not so. If you skip down to the 95% confidence interval, we are still rather certain the difference in means is not equal to zero, because zero is not in that interval either. But the 99% interval does contain zero. That tells us that our confidence is limited; we are rather sure that Company A produces LCDs with mean distance between recognition marks that are just slightly larger than those from Company B, but our confidence is not as high as 99%.

If we use a confidence interval to decide whether or not the means are different, we will have to come to terms with the acceptance of possible error that is inherent with the confidence interval. For judgments like this, there are more direct decision rules that can be constructed for this purpose. This idea is central to tests of hypotheses, which we discuss next.

## 4.2   TEST OF HYPOTHESIS

Like a confidence interval, a test of hypothesis helps us gauge uncertainty in an estimate. Unlike a confidence interval, this test will make a decision on whether the sample mean is large, small or sufficiently close to a target value. The outcome of the test can be simplified to a Yes/No answer, but like the confidence interval, it comes with caveats about the possible error in the decision. This simple outcome represents both a strength and weakness for the test of hypothesis. Compared to a confidence interval, the simple yes/no is easier to communicate to your audience when you are interpreting your data, but we will see below that assessing data in this framework is not always helpful if the hypotheses don't match your interests in the problem.

In the last section we were wondering if the mean distances for company A and B were equivalent. This is a conjecture or hypothesis that can be formed in a statistical test of hypothesis. To do this, we consider the two sides of the argument (the means are the same versus the means are different). Statisticians contrast the two points of view in two hypotheses: the *Null Hypothesis*, denoted by $H_0$, and an *Alternative Hypothesis*, denoted by $H_1$.

$H_0$ is called the null hypothesis because it stands for the "status quo", while $H_1$ is a conjecture that challenges the status quo. In most cases, if we want to see if two means are different, we arrange the hypotheses as

$H_0$: The two population means are the same

$H_1$: The two populations means are not equal.

Using this set-up, we allow the status quo to imply the means are the same, so our test of hypothesis is set up to show the alternative hypothesis (means are not equal) is actually true.

As a simple analogy, consider the prosecutor's job in a criminal court case. The legal maxim "innocent until proven guilty" implies that the case in which the defendant is not guilty is contained in the null hypothesis, which serves as the status quo. The onus is on the prosecution to uncover enough evidence to prove the alternative hypothesis, that the defendant is guilty.

### Interpreting Computer Output

Select both columns of data (for Company A and B) and choose

Hypothesis Test for 2 - Sample Mean

from the Analysis Menu. What you get is a comprehensive summary of all the possible conjectures we can impose between the means of two populations. First, the output is split into two parts, with the top half pertaining to the case in which we assume the two populations have an equal standard deviation (note that unlike the estimation procedure, we will not have the option to say the population standard deviations are known, so we must estimate them). The second half refers to a more general case, where the population variances can be different from each other. If given the choice, we hope to choose the top half.

In fact, in most comparisons of means from two different samples, the standard deviations are usually similar. If we assume equal variances for the LCD data, proceed to the next step of the two-sample analysis by deciding which of the three kind of hypothesis tests is appropriate for us. Here is the eZ SPC output:

```
1. Hypothesis    H0      mu1= mu2     mu1= mu2     mu1= mu2
                 H1      mu1>< mu2    mu1> mu2     mu1< mu2

2. Test statistics            t = 3.064283

3. Test result   p-value     0.002818     0.001409     0.998591
                 Result      Reject H0    Reject H0    Fail to reject H0
                 sign.level  5%           5%           5%
```

The output shows three columns pertaining three kinds of alternative hypotheses: $(H_1)$: mu1>< mu2, mu1> mu2, and mu1< mu2. Here, (mu1, mu2) = $(\mu_1, \mu_2)$ represent the means for the populations, and the symbol >< means "not equal". This is called a *two-sided test* and it is the hypothesis we are interested in testing, because we are curious about any difference between the two samples, and have no preconception that Company A would have a bigger mean than Company B except for what the data showed us (in general, forming your hypotheses about the data by first finding enigmas in the data set is "data snooping", and is considered unethical). So we are testing

$$H_0 : \mu_1 = \mu_2 \qquad \text{versus} \qquad H_1 : \mu_1 \neq \mu_2$$

and the decision will be based on how far apart the sample means are. The test statistic, from a two-sided t-test, is t = 3.064283, which tends to favor the null hypothesis if it is close to zero, but we won't worry about scrutinizing the t-statistic too much. Instead, to interpret the test statistic we look only at the *p - value*, which represents a summary statement about the apparent improbability regarding the null hypothesis.

## 4.3   THE P-VALUE

To understand the p-value, we first suppose the means are equal (so $H_0$ is true). This is similar to how a trial jury should first suppose the defendant is not guilty, and wait for evidence from the prosecutor before they change their minds. For comparisons of such two equal populations, some samples will differ in mean more than others, just by the pure chance and sampling error. At some point, if the difference between sample means is large enough, we need to decide that there is something more than pure chance involved, at which point we decide the assumption about null hypothesis is wrong, and we side with $H_1$.

The p-value is the probability of making a mistake using this tactic with our data. If the sample means are nearly equal, we should hesitate to side with $H_1$, and in this case the p-value will be somewhat large... generally anything over 0.20 is considered too high to switch allegiances from $H_0$. In our case, the p-value is

$$p = 0.002818,$$

which is considered small enough to confidently side with $H_1$, and decide with high confidence that the population means are different. The p-value can be tied in with this confidence; to be 95% confident in deciding the means are different, the p-value must be smaller than 0.05. We can make this statement about the LCDP data, but like the confidence intervals told us, we cannot use the 99% certainty level (the p-value is not less than 0.01). To simplify the testing procedures, you can use our general guideline on when to decide to buck the status quo of the null hypothesis and switch sides to $H_1$.

| p - value | Decision about null hypothesis |
|---|---|
| 0.00 – 0.01 | Very strong evidence favoring $H_1$. |
| 0.01 – 0.05 | Strong evidence, usually considered strong enough to decide $H_1$ is true. |
| 0.05 – 0.10 | Data supports $H_1$ somewhat strongly, but the evidence is in no way overwhelming. |
| 0.10 – 0.20 | Data suggest $H_1$ might be true, but but there is not enough evidence in the sample to state this with much confidence. |
| 0.20 – 0.80 | There is not strong evidence to support either hypothesis. |
| 0.81 – 0.99 | Data actually provide support for the status quo. |

We must point out one final note about the p-value before we go further. No matter how large the p-value, we have not suggested "accepting" the null hypothesis as true. Instead, we've stayed with a more general "do not reject" the null hypothesis. Remember, although the test can show support for $H_0$, it is not constructed to perform the task of proving $H_0$ is true. For example, if we set up a hypothesis test for a population mean, and sample contains only one observation, even if the observation seems strongly supportive of $H_1$, it will not constitute enough evidence to confidently reject $H_0$. On the other hand, given such an observation, it would be absurd to accept the null hypothesis as being true in this case. Frequently, an experimental outcome turns into a verdict of "do not reject" not because the data imply $H_0$ is true, but because the data did not provide sufficient evidence to prove it was false.

## Type I and Type II errors

The probability of accidentally rejecting the null hypothesis when it is actually true will never really be known unless you are doing simulations or validation tests. But this error is considered crucial to a test of hypothesis, and is called Type I error. Type II error refers to the complementary event, where we fail to reject the null hypothesis outright even though it was not true. Engineers and scientists often set up a statistical analysis to ensure the Type I error rate is less than 0.05. If we keep the Type I error rate down to 0.05, this means we can reject the null hypothesis in favor of the alternative if the p-value is less than 0.05. That is a general rule for statistical tests:

> **Reject the null hypothesis if the p-value**
> **is less than the allowed Type I error rate.**

In tests of hypotheses, there has been an emphasis on reducing the Type I error rate. But it is important to remember that two different kind of errors are possible, and the Type II error (failing to reject the null hypothesis when it is false) can also be consequential. When we try to reduce the Type I error rate in any fixed experiment, there is a tradeoff. Reducing the Type I error rate necessitates that we allow the Type II error rate to increase. Although we will not keep close tabs on quantifying Type II error in this lesson, it is good to keep in mind that if you want to improve your experiment by reducing the Type I error rate from 0.05 to 0.01, we have to sacrifice our ability to reject the null hypothesis when the alternative is true.

## Example 2-1: LCD Panel Quality Data

Let's go over another simple example with the LCD panel data. Select the first column of data and choose

Hypothesis Test for 1 - Sample Mean

from the Analysis Menu. This time, when you are asked to select variables, enter the number 25.22 (the LCD distance target value) into the box next to mu. Leave sigma unspecified, so the standard deviation will be estimated from the data. We are setting up a test where $H_0$ states the population mean is equal to the target value of 25.22. We are only interested in seeing if the mean has drifted away from 25.22, and a drift in either direction is important to detect. Because of that, we are interested in $H_1$ that states the mean is **not equal** to 25.22, which corresponds to the column in the output that states **mu >< 25.22**. The p-value in this case is 0.966948, which means the data in no way support $H_1$. A p-value larger than 0.5 is actually reinforcing the status quo of $H_0$.

$$\bullet \ \ \bullet \ \ \bullet$$

## One Sided Tests and Intervals

We have paid more attention to two-sided confidence intervals and two-sided tests of hypothesis. One sided tests are fairly common when looking for a process change in a particular direction.

For example, suppose the 50 LCD panels at Company B were scanned to see if any sample distance measurements were more than 0.005 units from the target value. Measurements that are within 0.005 are considered "high quality units", and we want to see if the percentage of high quality units is less than 90%. If so, there is a reliability problem within the plant and corrective action must be taken. In this case, we are interested in the event the proportion of high quality units ($q$) is less than 0.90, and this should dictate how we formulate the alternative hypothesis. The correct test would be based on the following:

$$q = \text{proportion of high quality units}$$

$$H_0 : q \geq 0.9 \qquad \text{versus} \qquad H_1 : q < 0.9.$$

In the sample of 50 total units, 42 of the units meet this standard, so our estimate of the proportion of high quality units produced by Company B is $42/50 = 0.84$, which is less than the 90% we were looking for. Should we take corrective action?

Not necessarily. By writing the hypotheses in this way, the status quo is to assume the quality is fine, and we placed the onus on data to prove there is a quality problem. We have chosen to take action only if the evidence strongly suggests it is needed. This is a test for attribute data. Select the column of data for Company B and choose

**Hypothesis Test for 1 - Sample Proportion**

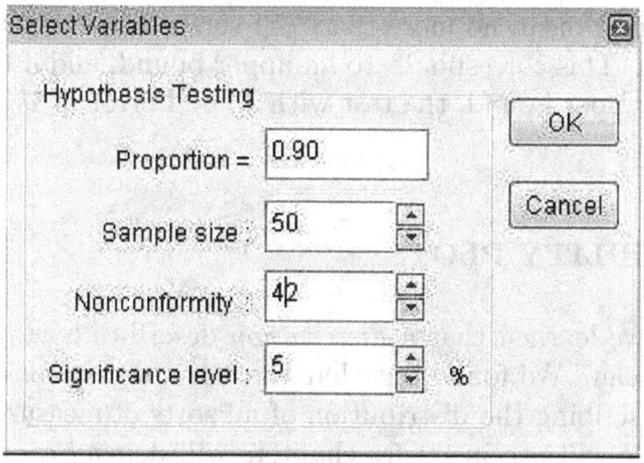

**Figure 4.1**    eZ SPC window for **Hypothesis Test for 1 - Sample Proportion.**

from the Analysis Menu. eZ SPC brings up the following window shown in Figure 4.1.

Enter the data as you see it in the figure and press OK. Your resulting output should be similar in style to the previous tests, but this time our eye is on $H_1 : q < 0.9$. For this example, the p-value was computed as 0.157299, which we know suggests $H_1$ might be true, but also acknowledges that there is not enough evidence in the sample to state this with a lot of confidence.

If we set our Type-I error rate to 0.05, we would not reject $H_0$ because the p-value is larger than 0.05. If we choose to make a confidence interval for the proportion of high quality units, the two-sided interval will give us a measure of uncertainty for our estimate of $42/50 = 0.84$. However, if we are only interested in knowing whether the proportion is too low (as we stated in the alternative hypothesis), then the kind of confidence interval we should use is not the typical two-sided *plus-or-minus* ($\pm$) interval. Instead, we want a lower bound for the proportion.

## Hypothesis Tests versus Confidence Intervals

We will wrap up this section by relating the two techniques we apply to describe uncertainty in our decisions. If someone presents results using a 95% confidence interval, will those results correlate with a test of hypothesis? Yes, it will. In fact, the 95% two-sided confidence interval corresponds to a test of hypothesis with a Type I error rate of 5%. For example, suppose you want to test to see if $H_0$: the population mean is 750 versus $H_1$: the population mean is not equal to 750. The number 750 is in the confidence interval if and only if the null hypothesis is not rejected. The 95% from the interval translates to the test of hypothesis in its type I error rate, which is $1.00 - 0.95 = 0.05$, or 5%.

One-sided tests can also be related to confidence intervals, but not the two-sided kind. Suppose we set up a one-sided test with type I error 0.01 based on

$H_0$: the population mean no more than 750 versus $H_1$: the population mean is greater to 750. This corresponds to an upper bound, and if the number 750 is less than the upper bound, the test with Type I error of 1% will not reject $H_0$.

## 4.4  PROBABILITY PLOTS

In Section 3.3, we learned that a *distribution* describes how data are spread out around a mean. We found out that the *normal distribution* is often appropriate for describing the distribution of all sorts of measurement data. In fact, methods we will use in future chapters will depend on the assumption that the normal distribution adequately describes the measurement data used in statistical process control. Assumptions, of course, do not always match reality. What do we do if the data are not described well by the normal distribution? How could we even tell?

One way we might judge the "normality" of the data is to inspect the histogram of the data and visually judge if the histogram is mound-shaped and emulates the actual probability distribution for the normal. If the data highly skewed, like data from a Poisson distribution (see plot (b) in Figure 3.5), you might be able to see this with your histogram. For most measurement data, however, this is a crud approach and will be insufficient. That is why we have a probability plot.

The *probability plot* provides us with a graph of the data points that have been re-scaled so that if the distribution of the data is normal, the graph will appear as a straight line. Unlike the histogram, it is much easier to see non-normality in the graph because the points will curve off the straight line on one or both sides.

Along with the probability plot, we can create a hypothesis test to determine if a set of data fits the assumptions of the normal distribution. Rather than test the mean or the variance of a data set, we can test the hypotheses

$$H_0 : \text{Data have a normal distribution} \quad vs.$$

$$H_1 : \text{Data have some non} - \text{normal distribution.}$$

This is sometimes called a test for *goodness of fit*. The way we will test this hypothesis is with the probability plot procedure, which is provided by eZ SPC. To show how it works, we will return to Example 2-1 (LCD Panel Quality Data). Recall there are two columns of data pertaining to distance between marks (in mm) for TCP bonded to LCD panels. If we open up `LCDP.ezs`, we can select the first column of data, which represents measurements from Company A. Under the Analysis menu tab, select

`Normality Test.`

You should see a window pop up as in Figure 4.2. Under **Parameter Selection**, the default choice is **Estimate Mean and Variance**. This is exactly what we want to do, so select OK to get your probability plot. The other choice of **Input** is only used when you want to test the hypothesis of a normal distribution with a specified mean and variance.

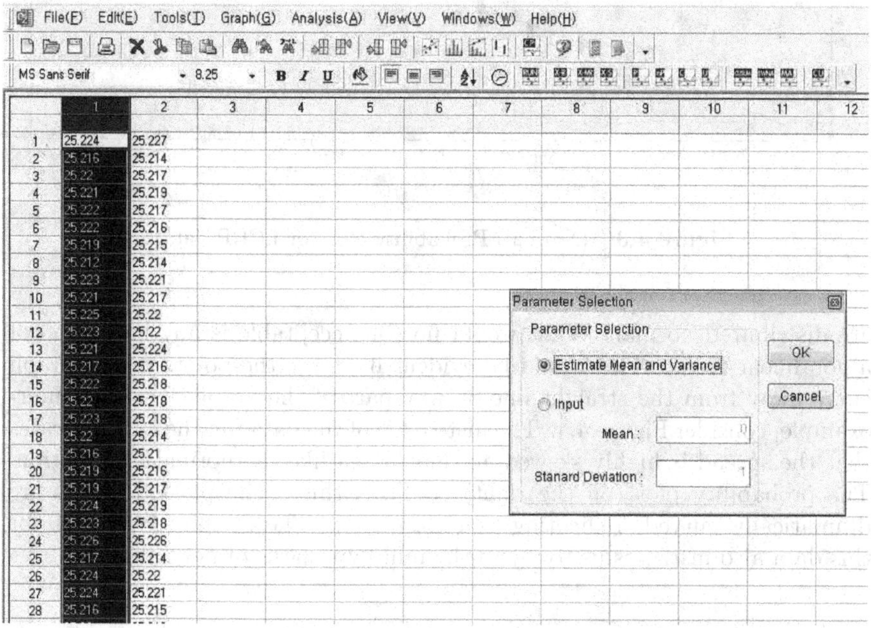

**Figure 4.2**   The eZ SPC menu command for probability plot.

The plot result you should get is displayed in Figure 4.3. A data summary is provided by eZ SPC as well:

```
Shapiro-Wilk Test Test Result
W=0.94093          Fail to Reject H0(=Normal Distribution)
```

We will not go over this brief and cryptic output except to say the hypothesis test for normality was conducted using a procedure known as the *Shapiro-Wilk Test*, and the end result was to conclude the null hypothesis of normally distributed data should not be rejected.

In this example, the data hover close by the straight line drawn through the graph, indicating the normal assumption is appropriate for this data. The corresponding histogram of the same data was presented in Figure 2.8 in Chapter 2. The straightness of the probability plot in Figure 4.3 is in agreement with the symmetry and "bell-shaped" appearance of the histogram.

We must emphasize, using this graph, that the plot does not have to line up perfectly straight. A little bit of nonconformity is expected if the data

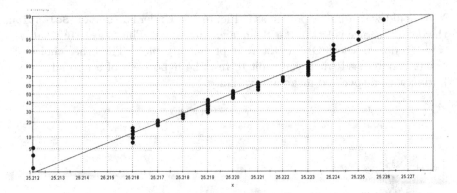

**Figure 4.3**    (Normal) Probability Plot for LCDP data.

are distributed normal. What would not be acceptable is any indication of a non-linear trend. This would be evident if we see the points on the graph curve away from the straight line on any part of the graph. For a counter-example, consider Figure 4.4. The histogram of the data (on the left) indicates that the spread is highly skewed and not at all like the normal distribution. The probability plot (on the right) confirms this. The plotted points are dramatically curved, indicating non-normality. The Shapiro-Wilk test for this data also instructs us to reject the null hypothesis of normality.

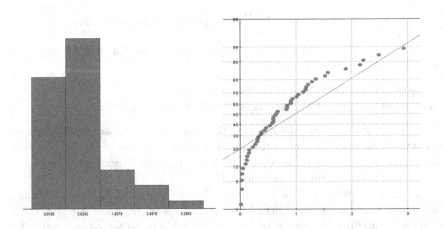

**Figure 4.4**    (a) Histogram of skewed data.  (b) Probability plot of same skewed data.

## 4.5   WHAT DID WE LEARN?

- Confidence intervals reveal uncertainty in a statistical estimate through their length and associated type I error.

- A statistical test of hypothesis partitions an experimental conjecture into a null hypothesis and an alternative hypothesis.

- The type I error of a test refers to the probability of rejecting the null hypothesis when it was actually true.

- The type II error of a test refers to the probability of not rejecting the null hypothesis when we should have rejected it.

- With an experiment based on a fixed number of sample observations, there is a tradeoff between Type I and Type II error.

- A probability plot is a graphical way of detecting whether the data is generated from a normal distribution.

- The p-value of a test is the smallest allowed type I error rate we could set that would still allow us to reject the null hypothesis.

## 4.6   TEST YOUR KNOWLEDGE

**1.** If your hypothesis test produces a test statistic with a p-value of 0.0819, which of the following actions are appropriate?

   a. If the Type I error is set to 0.10, we should reject the null hypothesis.
   b. If the Type I error is set to 0.10, we should not reject the null hypothesis.
   c. No matter what Type I error is considered, we should reject the null hypothesis
   d. Even if we know the Type I error rate, there is not enough information to decide whether or not to reject the null hypothesis.

**2.** Which of the following statements about the p-value is not correct?

   a. The p-value is probability of rejecting the alternative hypothesis.
   b. The p-value describes the conditional probability of rejecting the null hypothesis when it is known that the null hypothesis is actually true.
   c. We can reject the null hypothesis if the p-value is less than the Type I error rate.
   d. P-values larger than 0.20 do not adequately support the alternative hypothesis.

**3.** We want to show that Cow $A$ produces more milk than Cow $B$. On average, Cow $A$ produces $A_0$ liters of milk per day, and Cow $B$ produces $B_0$ liters per day. The daily production for each cow is recorded for 20 days. Which of the following hypothesis is appropriate for this test?

    a. $H_0$: $A_0 = B_0$ versus $H_1 : A_0 \neq B_0$

    b. $H_0 : A_0 \leq B_0$ versus $H_1 : A_0 > B_0$

    c. $H_0 : A_0 \geq B_0$ versus $H_1 : A_0 < B_0$

    d. $H_0 : A_0 \neq B_0$ versus $H_1 : A_0 = B_0$

**4.** Open LCDP.ezs and use eZ SPC to find a 90% confidence interval for the mean of the distance between recognition marks of the TCP for Company B (second column).

    a. (25.21710, 25.21874)

    b. (25.21727, 25.21904)

    c. (25.21741, 25.21890)

    d. (25.21815, 25.21941)

**5.** Again with Company B data, test the hypotheses $H_0$: mean distance is no more than 25.2 versus $H_1$: mean distance is bigger than 25.2. Enter the number 1 for the Significance level (the default is 5). This means we are performing a test with Type I error of 1%. What is the outcome?

    a. Reject $H_0$

    b. Fail to Reject $H_0$

    c. The test failed because p-value is stated as 0.000000

    d. The test failed because p-value is stated as 1.000000

**6.** Again with LCDP.ezs, select both columns (Company A and Company B), and choose Estimation 2 Sample Mean and then 2 Sample t-test from the Analysis tab. eZ SPC produces three confidence intervals (90%, 95%, 99%) for the *difference* of means between the two companies. Which of the three intervals contain zero?

    a. 99%

    b. 95%

    c. Both 95% and 90%

    d. None of the intervals

**7.** Again with LCDP.ezs, select both columns (Company A and Company B), and choose Hypothesis Test for 2 Sample Mean and then 2 Sample t-test from the Analysis tab. In the computer output (assuming equal variances), what is the p-value for the test $H_0 : \mu_1 = \mu_2$ versus $H_1 : \mu_1 \neq \mu_2$?

   a. 0.005218

   b. 0.002818

   c. 0.050000 and 90%

   d. 0.994782

**8.** If the experimenter decreases the Type I error rate from 0.10 down to 0.05, what happens to the Type II error rate?

   a. Stays the same

   b. Decreases

   c. Increases

   d. Can both increase or decrease

**9.** Suppose we set up a one-sided test with type I error 0.01 based on $H_0 : \mu \geq 3000$ versus $H_1 : \mu < 3000$. This corresponds to a lower bound confidence interval, and if the number 3000 is more than the calculated upper bound, the test with type I error of 0.01 will

   a. Reject $H_0$

   b. Not reject $H_0$

   c. Remain undecided about $H_0$

   d. Will randomly reject or accept $H_0$.

**10.** A normal probability plot will detect non-normality in the data set if the plot is

   a. curved

   b. straight

   c. equally spaced

   d. binomial

## EXERCISES

**4.1** A soft drink bottling company produces 355 ml cans. After soft drink is inserted into the can, its target weight is 390 grams. From long history of production, can weight is assumed to follow normal distribution. 50 cans were randomly selected and weighed. The data (in grams) are given in `Bottling.ezs`. Construct a 95% confidence interval for the mean weight. If a can contains over the allotted 395 g, there will be a risk of a soda explosion. Should the bottling company be worried?

**4.2** Cell phone battery life is one of important considerations when choosing a cell phone. The talk-time battery life was tested for 30 cell phones of the same model. Assume that this battery life follows normal distribution. Find the 90% confidence interval for the *variance* of talk-time battery life of this cell phone model.The data, listed in hours, are found in `CellPhone.ezs`.

**4.3** The marketing solutions manager of an email provider wants to improve her company's email deliverability rate. The company recently sent 236,138 emails on behalf of its users and 224,986 emails were successfully delivered to customer in-boxes. Find the 95 percent confidence interval of the true email deliverability rate.

**4.4** A 1986-era 18 c.f. refrigerator uses 1400 kWh a year, while a modern energy-efficient model uses only 350 kWh. A refrigerator manufacturer claims that their new model uses on average only 300 kWh. To test this claim, 40 refrigerators were randomly selected and tested for energy consumption. The data are found in `Refridgerator.ezs`. Assuming that the kWh measurements follow a normal distribution, test the manufacturer's claim about the mean energy consumption at the significance level 0.05. What did you use for the null and alternative hypothesis?

**4.5** A trucking company provides truckload carrier services for retail merchandise, foods, and beverages throughout seven South East states in the United States. The company is considering a new route that might reduce the driving time over the existing route. In order to make sure new route saves time, the company had 20 trucks run each route to the same destination, respectively. The data, in hours, are given in `Trucking.ezs`. Assume that the driving times follow a normal distribution. Test to see if the new route saves driving time over the existing route at the significance level 0.05.

**4.6**   Open the file `ProbabilityPlot.ezs` in the Applications Folder. There are data in 14 columns and 50 rows. Each column is meant to represent a different sample. The goal of this exercise is to test whether or not each sample (column) has a normal distribution. To do this, select any column and, then under the Analysis Tab, select `Normality Test`. There are two pages of output produced by the normality test: The (Normal) probability plot and the Shapiro-Wilk test result.

Repeat the normality test for each of one of the 14 columns so you can compare plots. Some of them are show points plotted in a close line, but other plots show various forms of *nonlinearity* in the form of curved lines, bumps in the line, and outlying points. Use your eyes and gut feeling to decide whether the plot is straight enough to imply the underlying distribution of the data is Normal. Which probability plot is the straightest? That would be the column of numbers that look most like Normal data. Which ones are the most curved?

**4.7**   This exercise is a continuation of the last one. After you have mulled over the probability plots, you can use the Shapiro-Wilk test statistic to check your results. On the right of the test statistic, eZ SPC will tell you if the data look strongly non-Normal by the Test Result saying `Reject H0(=Normal Distribution)`. There are two plots that are highly nonlinear and the null hypothesis of normality is rejected.

You can use the Shapiro-Wilk test statistic to order the plots in terms of how Normal the data appear. For data that looks like it is from the Normal distribution, the statistic will be between 0.98 and 1.00. As the Shapiro-Wilk test statistic decreases, the data appear less normal. In some cases, you might have detected curvature (a sign of non-Normality) once the test statistic decreased to 0.96. Your instinct might be right. However, the test result is inconclusive, so the null hypothesis (of Normality in the data) was not rejected.

# CHAPTER 5

---

# SHEWHART CONTROL CHARTS

---

In Chapter 1, we mentioned that after completing the lessons in this book, you will have at your disposal a *tool box* full of tools and techniques to help with quality improvement. The control chart ranks first among these tools, which also include Pareto charts, flow charts, histograms, cause-and-effect diagrams and other visual aids. Control charts are used to measure and understand the variability of a process. By reducing this variability, we achieve quality improvement. If the observed process variability can be traced to an assignable cause, our goal is to identify and remove that cause as soon as possible. In this chapter, we are going to explain how to interpret the process this way by using control charts.

## 5.1  THE CONCEPT OF A CONTROL CHART

Figure 5.1 illustrates a simple process with natural variation. As the timeline goes from lower left to upper right, process measurements are made at six time points, and the timeline itself represents a target value for the process. At each one of these time points, four samples of the process output are measured, so a total of 24 measurements are taken from start to finish. The

*Basic Statistical Tools for Improving Quality.* By Chang W. Kang and Paul H. Kvam  **97**
Copyright © 2011 John Wiley & Sons, Inc.

bell-shaped curves represent the (normal) distribution of the data and the apparent variability within each cluster of four measurements. Notice that the curves have approximately the same shape and spread, indicating that the variance within each group of four clusters is about the same. Also note that the location of the middle of the bell curve changes at different time points; the mean value of the first cluster appears close to the target line, but second cluster less so.

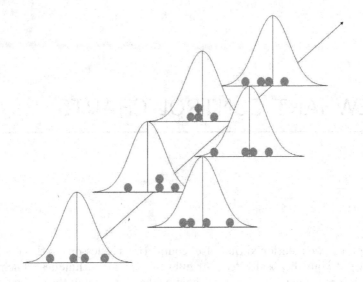

**Figure 5.1**   Variation in a process where 4 samples are drawn at each of 6 time measurements.

This is the basis for a control chart, which is used to monitor variability in a process and identify the assignable cause if it exists somewhere in the process. If an assignable cause of variability is not detected as early as possible, the process may produce scores of nonconforming or defective products. The basic control chart, shown in Figure 5.2, is made up of three horizontally parallel lines: the *centerline* in the middle, representing the output target, and two control limits. The one above is called the *upper control limit* (UCL), and one below is the *lower control limit* (LCL). The process is deemed to be in control if the output values are within these control limits.

Warning to the reader: it is easy to confuse control limits with specification limits. The *lower specification limit* (LSL) and *upper specification limit* (USL) are used to determine acceptance of a product. The control limits, on the other hand, are used to determine whether the process is in an 'in control' state or 'out of control' state.

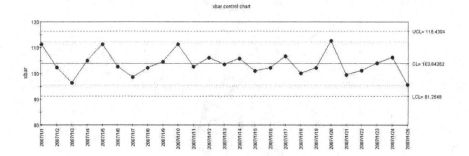

**Figure 5.2**    Example of a Control Chart.

By placing the UCL and LCL at a particular distance away from the centerline, we have actually implied a specific amount of allowable tolerance for variation in the process. The further away from the centerline we place the UCL and LCL, the more tolerant we are to output values that stray from the intended target. If they are too far from the centerline, we lose the ability to discover assignable-cause variation because the control chart will fail to make us aware of it. By putting the UCL and LCL too close to the target, the natural variation in the system can cause us to believe the process is out-of-control even if no assignable cause of variation is found. The trick is to find just the right place to place these boundaries on the control chart.

The natural variation in the control chart pictured in Figure 5.1 shows that the output values from each time point are distributed around a mean output value and have the same approximate variance or standard deviation (see Chapter 3), which we denote with the Greek letter $\sigma$. Shewhart used $3\sigma$ control limits in his original control charts. A $3\sigma$ control limit means that the upper control limit is set at three standard deviations (or $3\sigma$) above the centerline and the lower control limit is set at three standard deviations below the centerline. Assuming that the process follows the normal distribution (implied by the bell-shaped curve shown in Figure 5.3), the probability that a sample output stays between the LCL (3 standard deviations less than the process mean) and the UCL (3 standard deviations above the process mean) is 0.9973. In other words, the probability of being outside the control limits is a fraction of one percent: $1 - 0.9973 = 0.0027$.

If the sample data stay within the control limits, the process is considered to be statistically in control (simply "in control") since we expect 99.73% of sample data from stable processes will stay within these control limits. If a sample output goes above the UCL or below the LCL, the process is considered to be *out-of-control*. That is, we choose to believe that previous assumptions about the process being in control are now wrong, and we reject the notion that the process is in control and that it just happened to produce such an unlikely output. At this $3\sigma$ limit, we no longer recognize the variability we

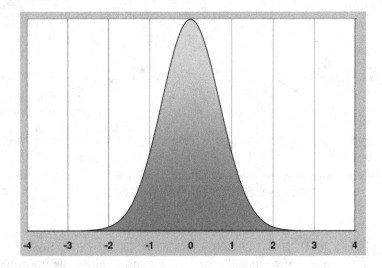

**Figure 5.3**   The bell curve of a normal distribution of output values with $\sigma$-unit spacings from $-4$ to $4$.

are observing as a natural part of the process. The probability 0.0027 can be related to the type I error rate in the test of hypothesis described in Chapter 4 (Analyzing Data). Under the null hypothesis, we believe the process is in control. If we reject the null hypothesis according to the $3\sigma$ rule, then the process has a type I error rate of 0.0027, representing the probability we mistakenly claim the process is out-of-control when it actual is stable.

One of the fundamental challenges of managing a business process is discerning the natural variation of the output from assignable-cause variation. When the process is in control, we are inclined to believe the differences in the process outputs are due to chance cause. On the other hand, if the process is deemed out-of-control, we are more apt to believe that some of the process variation is due to an assignable cause. Since the variability in the out-of-control process is seemingly beyond our comfort zone, the future behavior of process will undoubtedly be too erratic and unpredictable. In the battle against process instability, the control chart is our first line of defense.

## 5.2   MANAGING THE PROCESS WITH CONTROL CHARTS

As the process manager, you want to improve the process by reducing its variability. By reducing the process variability, you can minimize the quality cost and maximize customer's satisfaction. The control chart is an excellent tool to monitor the process, identify assignable causes in the process, and direct the process. In order to construct a control chart we need to decide control limits, sampling size, and sampling interval for effective process control.

## Choice of Control Limits

The control limits are set to determine whether the process is in control or not. Many of today's manufacturing processes use Shewhart's $3\sigma$ control chart, and this serves as a good "default" choice. As mentioned earlier, the type I error rate for the three sigma control limits is 0.0027. Because the control limits depend on the value of process standard deviation, calculation of process standard deviation is very important. Depending on the number of observations in the sample, we need statistical knowledge for calculating standard deviation but the included software eZ SPC will provide the control limits.

Since the objective of a control chart is to detect an assignable cause of variation as early as possible, the use of *warning limits* is sometimes suggested. Warning limits are similar to the control limits, but generally placed closer to the center line to alert the process manager more quickly in case outputs are straying unusually far away from the target mean, but not yet past the control limits. The purpose of warning limits is to send a red flag to the process manager without signaling the process is out of control, as the regular control chart limits are set to do. Usually the warning limits for $3\sigma$ control charts are set at two standard deviations ($2\sigma$) above and below from the center line.

## Choice of Sample Size

In the simple process plotted in Figure 5.1, four measurements were averaged at every time point. By taking multiple measurements, we are allowed to observe the variability of outputs at that particular time point, and this information about the process is invaluable; if we use larger samples we are able to detect smaller shifts in the process. Of course, we are limited by budget and time constraints regarding how many measurements we sample. In some processes, it is physically possible to sample only one measurement at a time. In any case, after sampling a certain amount of data, the expense of sampling will eventually outweigh the benefits. In a well-organized process, the sample size we choose at each time point must be specified before we construct a control chart.

When the process seems quite stable, it may not be necessary to take a large number of samples at each time point, so we can save our resources by sampling less. However, if the process has the potential to be unstable then we need to take more measurements in order to accurately detect any shift in the process. If we are without expert knowledge about the cost of sampling, a rule of thumb says we should sample 4 or 5 output measurements at each time point.

### 5.2.1    Choice of Sampling Interval

So far, we have discussed how many observations should be measured at each time point, but not what those time points should be. Every hour? Every day? Like the sample size, the length of the sampling interval must be specified before we construct a control chart. Naturally, this depends on how the process is perceived to change over time. Certain processes change over months and others change by the second. If the process seems stable, we should refrain from interrupting it by repeatedly sampling after short intervals of time. As another simple rule of thumb, consider a potential time interval (time between collecting output measurements) and ask yourself if the interval is short enough so that you are highly confident that nothing unusual could have happened to the process between the times you sampled. That is, you should be able to assume the process drifting out-of-control and back to in control in this time span to be an unrealistic concern.

Like the choice of sample size, our chosen sampling interval length is also a function of cost. In applications where the sampled output can be damaged or destroyed, production and repair costs are chief concerns. Aside from the cost of interrupting the system to take measurements, there is also potential cost in the act of sampling. Unfortunately, the value of this kind of analysis can be diminished if the cost of sampling is so high that the process manager is essentially forced to choose a sampling interval that is much longer than the preferred one.

### Performance of Control Chart

The number of points plotted on the control chart before it indicates an out-of-control condition is called the *run length*. The average number of run lengths you observe from a process is called the *average run length* (*ARL*) and is used to evaluate the performance of control charts. The *ARL* is a favorite statistic in statistical process control, and it is easy to interpret. If the process is in control, we denote the average run length as $ARL_0$ and $ARL_1$ if the process is out-of-control. Obviously, $ARL_0$ should be a large number, indicating that the process is expected to produce consistently conforming products before producing any defective products that cause an out-of-control signal to the process manager. If the process is out-of-control then a small $ARL_1$ value is a signal that more investigation is needed. Recall Shewhart's $3\sigma$ control chart, which has 0.0027 as false alarm rate. In other words, when the process is in control, the probability that any point exceeds the control limits is 0.0027. If the measurements are *uncorrelated* (that is, measurements are not directly influenced by previous measurement values), then $ARL_0$ can be calculated as

$$ARL_0 = \frac{1}{0.0027} = 370.$$

This means that even though the process remains in control, an errant out-of-control signal will be generated once every 370 samples on the average.

If the process is even slightly out of control, we expect a smaller average run length. When monitoring the process mean, if the mean shifts higher or lower than the target, a $3\sigma$ chart should go out of control more frequently than once every 370 times. If the mean increases in magnitude of one standard deviation, for example, the average run length ($ARL_1$) for an $\bar{x}$-chart based on $n = 4$ measurements per sampling period (described in the next section) is 6.3.

In the next sections, we will look at the different kinds of control charts that are available to chart a process. They can be divided into two categories: *variable control charts* and *attribute control charts*. As a process manager, you will be able to select the appropriate control chart for your process and set the parameters that determine how sensitive the chart will be to outputs that stray from the target value. When we select these parameters of the control chart, such as its control limits, sample size, or sampling frequency, we can consider them in regard to $ARL_0$ and $ARL_1$ to gauge the performance of control chart. For a process that is in control, we can increase or decrease the $ARL_0$ by changing these input parameters.

## 5.3   VARIABLE CONTROL CHARTS

A quality characteristic of the process output which is measured on a numerical scale is called a *variable*. Examples of a variable are thickness, weight, length, volume, strength, and concentration. Variable control charts are used to evaluate variation in a process where the measurement is a variable. When we deal with variability in a measured quality characteristic, it is typical to monitor both the sample mean value of the quality characteristic and its sample variance. To measure the mean, we use the sample average ($\bar{x}$), and to consider the variability, we consider both the sample standard deviation (s) and an even more basic measure of dispersion, the sample range (R), which is the difference between the biggest value and the smallest value in each sample. Although the sample standard deviation (Chapter 3) is a better statistic for measuring sample variability, its performance is not so great with samples less than size 10. That is why the more rudimentary R statistic is considered with small samples.

To monitor the quality characteristic according to both mean and variability, we introduce three kinds of control charts: $\bar{x}$-R control charts, $\bar{x}$-s control charts, and x-MR control charts.

### $\bar{x}$-R Control Chart

By plotting the sample mean ($\bar{x}$) in a control chart, we can monitor the process to see if the mean output levels are consistent or if they are drifting away from the target value. By plotting the R statistic in a control chart (recommended when the sample size is less than ten), we can monitor the process to see if variability in the output remains consistent across samples. If these two procedures work together in concert, we can monitor both at the same time.

In order to construct an $\bar{x}$-R control chart, we need to estimate $\sigma$, the standard deviation of the process, and the range provides our best guess of the unknown standard deviation when the sample size is relatively small (if the range is applied with a larger sample, the computer software will use different multipliers of the range statistic to estimate the standard deviation, depending on the sample size). We can construct the $\bar{x}$ control chart and R control chart separately for the quality characteristic of interest, but we will use both control charts at the same time to judge whether the process is in control. We take an example to illustrate how to construct $\bar{x}$-R control charts and interpret the results.

## Example 5-1: Tissue Paper Strength

A consumer product manufacturing company produces tissue paper. We want to inspect the tissue cutting process and decide whether the process is in statistical control by inspecting machine direction tensile strength (MDTS) of the tissues. The MDTS of the tissues is required to be resistant to breaking when the raw paper is cut. The measuring unit for the machine direction tensile strength is actually grams/76mm-2plies. The target value is 937 grams/76mm-2plies and the specification limits are as follows: USL: 1060 grams/76mm-2plies and LSL: 825 grams/76mm-2plies. Twenty five samples, each of size five, have been taken. The data are can be found in the file MDTS.ezp in your applications folder. Follow these steps in sequence to construct an $\bar{x}$-R control chart for the tissue paper data:

1. Open the file MDTS.ezs in your applications folder and select the five columns of data.

2. From the Graph menu bar, select the $\bar{x}$-R control chart from the pull-down menu.

3. We are using the data for process analysis (rather than process control), so click OK for this default choice.

4. Click OK again for another default: X-axis based on the order of the data.

5. eZ SPC will generate the $\bar{x}$-R control charts at the same time. The control chart on top is the chart and the one on the bottom is the R chart.

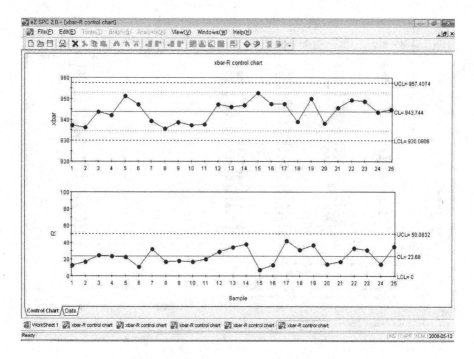

**Figure 5.4**   $\bar{x}$-R chart for tissue MDTS.

The control chart in Figure 5.4 can be printed, copied or saved. As an option, you can copy the graphs for the $\bar{x}$ control chart and R control chart separately. You might notice that eZ SPC actually produced two windows of output in this procedure. If you select the Data tab on the bottom left side of the window, you will see computed values of UCL, CL, and LCL for both the mean (from the $\bar{x}$ chart) and the range (from the R chart). The data sheet is shown in Figure 5.5. Notice the extra columns labeled $\bar{x}$ Code and R Code. Entries in these columns inform the user of any departure(s) from the in-control state.

If the process appears out-of-control according to the $\bar{x}$-R chart, the information on the Data tab provides vital information for finding the assignable-cause variation and planning process improvement. Along with detailed statistics on the mean and range statistics found on this page, it also contains remarks about the process that may be of interest to the user. Any values that strayed outside the control limits are marked with an asterisk (no values are marked in this example), and the notes also point out small anomalies with the charted data that might be of interest, including unexpected runs. eZ SPC lists nine comments (A through I) about the number of runs found in different regions of the control chart:

A   seven successive points are on the same side of the centerline

B   seven or more successive points tend to increase or decrease

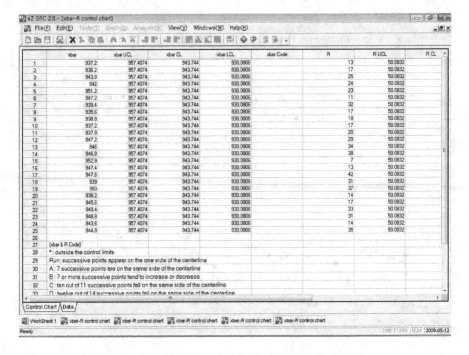

**Figure 5.5**  Data sheet provided by eZ SPC for $\bar{x}$-R chart.

C   ten out of 11 successive points fall on the same side of the centerline

D   twelve out of 14 successive points fall on the same side of the centerline

E   fourteen out of 17 successive points fall on the same side of the centerline

F   sixteen out of 20 successive points fall on the same side of the centerline

G   at least two out of 3 successive points fall in between $2\sigma$ and $3\sigma$

H   at least three out of 7 successive points fall in between $2\sigma$ and $3\sigma$

I   at least four out of 10 successive points fall in between $2\sigma$ and $3\sigma$

If any of the events from A to I occur in the control chart, the software will produce output to alert the user in a special column titled "out-of-control code". Consider these nine comments as helpful or insightful observations and not warning sirens of impending doom. When we check the details of the plotted points on the control chart, we hope to see a random pattern to indicate that the process is in control. In this case, a "run" was defined as successive points that appear on the one side of the centerline. If the pattern is truly random, each successive observation has a half chance of falling above or below this line, according to the rules of probability. That means run lengths average around 2 and seldom grow larger than 5. In this example, there were eight instances where the chart exhibited a run that would be considered improbable under this assumption of complete randomness.

To a statistician, these numbers only represent irregularity. To someone who thoroughly understands the process, these numbers might represent helpful clues for finding potential sources of assignable-cause variation, even in cases the process was considered to be stable. However, if no further inquiry is warranted, then we make note that all the output measurements from the chart are contained within two standard deviations, suggesting the process is in control with regard to change of output mean.

The R chart plots only the UCL to go along with the centerline, and the LCL is set to zero because the LCL for an R control chart cannot be negative. In this example, no observed R statistics fall beyond of the computed UCL limit, suggesting the process is in control with regard to having consistent variance.

$$\bullet \ \bullet \ \bullet$$

### 5.3.1  $\bar{x}$-s Control Chart

Our rule-of-thumb is to use the $\bar{x}$-R chart if sample sizes are less than 10. Once we get a sample of size 10, we suggest the $\bar{x}$-s chart, which replaces the chart of the sample range with one of the sample variance, keeping the $\bar{x}$ chart the same. Since many of the steps in constructing the $\bar{x}$-s chart are the same as for the $\bar{x}$-R chart, we won't repeat them with the same detail. Instead, we focus on the s chart, and how it differs from the R chart.

### Example 5-2: LCD Brightness

The luminosity, or brightness measurement, of a LCD monitor is an important quality characteristic. Low or unsteady brightness indicates poor quality, so LCD manufacturers keep track of monitor brightness of the monitors during the developmental and testing process. Our example features 12 sample units of a brand of 17" TFT-LCD panel monitor. Each of the twelve units is tested and display brightness is measured in cd/(meter square), where cd is unit of candela (luminous intensity). The specification limits of brightness are between 150 and 200 cd/(meter square). The data are found in the file LCD.ezp in your applications folder. The sequence of steps for constructing $\bar{x}$-s control charts is about same as those for constructing the $\bar{x}$-R control charts except for calculating the sample standard deviation for each sample. The steps are:

1. Open the file LCD.ezs in your applications folder and select the five columns of data.

2. From the Graph menu bar, select the $\bar{x}$-s control chart from the pull-down menu.

3. Click OK for the default choice of process analysis.

4. Click OK for default X-axis, based on the order of the data.

5. eZ SPC will generate the $\bar{x}$-s control charts at the same time.

According to the $\bar{x}$-s control charts in Figure 5.6, this process appears to be in control. By reviewing the information from the Data tab, we observed no averages or standard deviations outside their respective control limits. eZ SPC lists nine features of the charts that might be of interest to the process manager. Mostly, these are unusual runs of values above or below the center line, hinting at the possibility that sequential measurements could be correlated. Unless the process manager can find an assignable cause for this correlation, we can consider such runs as inconsequential.

In this example, the sample sizes remain constant (ten columns of data representing ten samples at each time measurement). There is no trouble if we have a variable sample size, but we will produce a chart with changing UCL and LCL, depending on the sample size. In general, equal sample sizes are always preferred.

Traditionally, quality engineers have favored the R chart over the s chart for the sake of computational simplicity. With the help of computer software such as eZ SPC, the computational difficulty has been eliminated and the $\bar{x}$-s control charts can be constructed with a couple clicks of the mouse. Remember the rule of thumb; if your sample sizes are less than 10, choose the $\bar{x}$-R control chart. If sample sizes are 10 or more, you will get improved results by using the $\bar{x}$-s control charts.

• • •

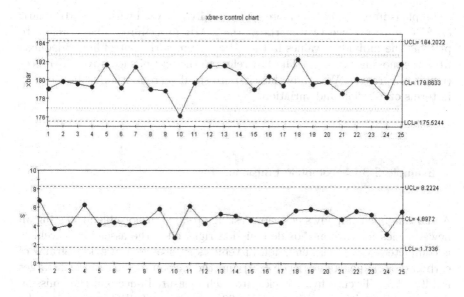

**Figure 5.6**   $\bar{x}$-s chart for LCD monitors.

## x-MR Control Chart

In some applications, it is impossible to meet the sampling requirements of an $\bar{x}$-R or $\bar{x}$-s chart. There are many situations in which we can only sample once for monitoring and controlling the process. That pretty much eliminates any chance of making an R-chart or an S-chart. As an alternative, the *moving range* (x-MR) control chart is a simple way to show how the process is changing over time.

The first chart is just a basic chart of the measured values across time - essentially a $\bar{x}$ chart with a sample size of 1. In some books, this is called an *individuals chart*. With only one observation at each time point, we cannot use sample variance or range to figure out the control limits. Instead, the x-MR chart estimates variability in two steps, first computing all of the differences between successive observations. We use the *absolute value* of the differences, because we are only interested in the distance between two successive measurements and not whether the first one is bigger than the second one. This list of differences between successive observations is called a *moving range*. In the second step, the x-MR chart uses the average value from the moving range data as an estimate for standard deviation ($\sigma$).

There are many potential applications where an $\bar{x}$-R chart or an $\bar{x}$-s chart is not possible, so an x-MR chart should be used instead. For example, when an automobile company tests vehicles for carbon dioxide emission, there is one measurement for one car, and x-MR control charts are useful for monitoring and controlling such a process. Any process for which the cost of measuring a sample is immense can be more economically analyzed with an x-MR chart. eZ SPC will produce two charts in the x-MR procedure. The X refers to plotting the individual values in time succession, with control limits deduced from the moving averages. The MR refers to the plot of successive differences (moving averages). We study this chart to see if there is stability in the plot, in terms of location and variation.

## Example 5-3:  Automobile Emissions Test

An automobile company's emission testing group is responsible for testing all new vehicles for carbon dioxide emission right after the assembly of vehicle is completed. $CO_2$ mass emission in vehicles is measured in g/km (grams of carbon dioxide per kilometer driven) and the carbon dioxide emission target is 140 g/km. Twenty five vehicles are each measured once and the emission data (in g/km) are listed in the file CO2.ezs. The eZ SPC software can be used to construct the x-MR control charts with following steps:

1. Open the file CO2.ezs in your applications folder and select the column of data.

2. From the Graph menu bar, select x-MR control chart from the pull-down menu.

3. Click OK for the default choice of process analysis.

4. Click OK for default X-axis, based on the order of the data.

5. eZ SPC produces both control charts ($x$ on top, MR on bottom).

From the x-MR control charts in Figure 5.7, this process is in control. The $x$ values never cross the UCL (which is of chief concern in this example). The MR chart looks jumpy and sporadic, as it should when the process is in control. We do not see any long runs of large or small differences, and the process appears stable.

• • •

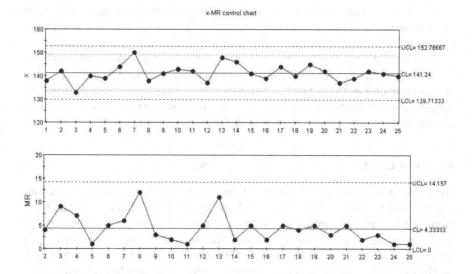

**Figure 5.7**   x-MR chart for auto emission data.

## 5.4   ATTRIBUTE CONTROL CHARTS

The variable control chart from the last section plots a quality characteristic in the process output that can be measured numerically. For many processes, output can only be classified as conforming or nonconforming. It is important here to define what we mean by defect (or nonconformity) clearly. A *defect* is a single nonconforming quality characteristic. Defective items are those items having one or more *nonconformities*. A quality characteristic that is classified into either *conforming* or *nonconforming* according to its specification is called an attribute.

Attribute data are *qualitative data* that can be counted for recording and analysis. To the statistician, this is lower quality data, since variable data contains potentially more information than just the attributes of the output. For example, if the trace amount of lead in a plastic toy is measured for each product output, we are measuring a variable quality characteristic. If the process only detects whether the amount of lead in the plastic toy exceeds a threshold value, then we are collecting attribute data. From the variable data, we could easily decide the product's attribute (exceeds threshold versus does not exceed threshold), but we can not retrieve the variable data when we only have the attributes.

As another example, attribute data might be the number of nonconforming computer chips on a wafer. In a service industry process, "pass" might mean the customer made it completely through the process, and "fail" means that the customer did not complete the process/sale. The analysis of attribute data was first discussed in Chapter 2. In the example from that section, 50 LCD panels were classified as conforming if sample distance measurements were less than 0.005 units from the target value. Instructions on graphing qualitative data using eZ SPC are given in Chapter 2. To chart attribute data, we will discuss four different control charts based on counts (number of observed non conformities) or proportions of counts (number of defective outputs divided by the total number of outputs surveyed). The four charts are the p-chart, the np-chart, the c-chart and the u-chart.

## p Control Chart

For a batch of items to be inspected on a pass/fail basis, the nonconforming fraction is the number of nonconforming products divided by the total number of inspected products. When we are interested in monitoring and controlling the nonconforming fraction, the p-control chart is the appropriate choice. Here, p is the fraction of the population that is nonconforming. At the different inspection times, when the nonconforming fractions are computed from their respective samples, it is common to experience differences in the observed sample sizes. This is technically alright, but this will lead to constructing control limits that change, depending on the sample size.

The nonconforming fraction changes from sample to sample due to the natural randomness inherent in the selection process. Figure 5.8 shows the distribution of this fraction when 25% of the overall product is considered nonconforming and ten items are sampled. With ten observations, the nonconforming fraction will be 0%, 10%, 20%, and so on, up to 100%. We expect 20% or 30% from a sample of ten will be nonconforming. From Figure 5.8, we can see it is plausible to observe any nonconforming fraction between 0% and 60% with a sample of ten. The count data is said to be distributed as *binomial*, which was introduced in section on statistical distributions in Chapter 3. For the standard deviation of defects from this distribution, the computed $3\sigma$ control limits can be implemented as before, so the basic veneer of the p-chart is the same as the previously mentioned variable control charts.

## Example 5-4: Video Game Quality

An electronics manufacturer produces computer video games that are played on a personal computer. The company's quality engineer inspects randomly selected samples of computer games off the assembly line every day, determines if any of them are nonconforming in any way, and gauges the control of the manufacturing process by the constructing a p chart. The data are listed in the file Games.ezs. The p control chart by eZ SPC 2.0 is constructed in Figure

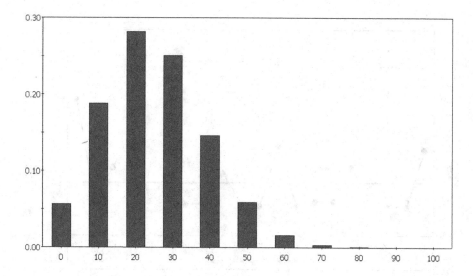

**Figure 5.8**    Distribution of the nonconforming fraction sampled from random lots of size 10 when the proportion of defectives in population is 25%. The horizontal axis is plotted in percentage (from 0% to 100%).

5.9 using similar steps that were used with variable control charts. The output from Figure 5.9 shows that the daily nonconforming fraction never exceeds the UCL, so the process is in control. Note how the control limits change because the number of games inspected on each day is not constant. In many examples like this one, the UCL is the key to the chart, and the LCL has no directly helpful purpose in the quality control process.

• • •

## np Control Chart

When we are interested in the *number* of nonconforming products rather than the nonconforming fraction, the np control chart is preferred instead of the p chart. For the np control chart, the size of the inspected samples at each time point must be the same, otherwise the comparisons of the number of nonconforming products are invalid. In a practical sense, the np chart and the p chart are doing the same thing, but if sample sizes are the same, the np chart is usually preferred because of the ease of interpretation. With many observed sample sizes, process managers will more easily follow the change in the number of nonconforming items observed at each time point rather than the non conforming fraction. For example, if n = 13, it will be easier to track

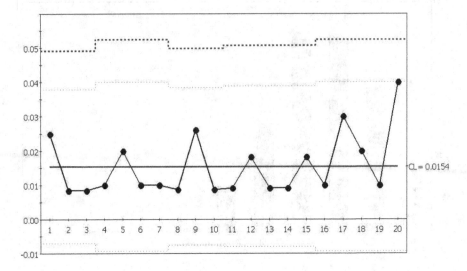

**Figure 5.9**    p chart for computer game inspection data.

the number of nonconformities (0, 1, 2, 3, ...) compared to the observed proportion (0%, 7.69%, 15.38%, 23.08%, ...).

## Example 5-5: Golf Club Shafts

A golf club shaft manufacturing company inspects one sample size of 100 everyday. The data for nonconforming shafts are given in the file Shafts.ezs. The np control chart is given below. In constructing the np chart, eZ SPC requires only one column of data input; the software will prompt the user for the total sample size in a separate window.

As before, the UCL is of chief concern. Although the process, as plotted in Figure 5.10, appears to be in control, there is an unusual characteristic of the plot that your eye might not pick up on right away. Starting with the sixth test, the test results show the number of defects oscillates between a small number of defects (one or two) and a relatively larger number (three or more). The process manager should be interested in this apparent correlation between successive tests. For example, what if the product testers alternate on successive days? Then it would appear the product tester influences the test result. If no such assignable cause can be identified, we will have to regard this anomaly as random variation.

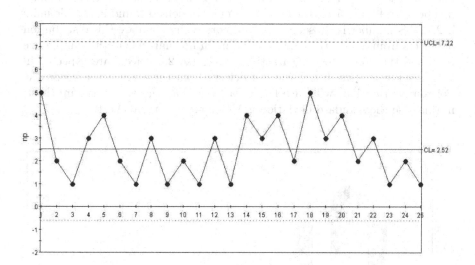

**Figure 5.10**    np chart for golf shaft measurements.

## c Control Chart and u Control Chart

For count data based on picking out defects from a finite lot, the np chart was recommended. In some applications, we are counting things, but they are not from a fixed number of inspected products. For example, in a textile mill, the swatches of fabric are continuously produced. In order to detect the flaws as early as possible, a 5-meter swatch of fabric is sampled and inspected by a skillful inspector. Fabric producers aim to control the number of flaws in the sample fabric. Unlike the np chart, where the number of defects in a lot is limited to the sample size, the defect count in this scenario is not limited, and there is no sense of what "proportion" would mean in this case. When we are interested in monitoring and controlling this type of attribute data, the Counts control chart (or c chart) is appropriate.

If the size of the unit differs from sample to sample, the expected number of counts per unit will also naturally change. For example, the number of defects found on a square foot of an automobile's painted surface would be four times larger than the number of defects found on a 6 inch by 6 inch square, since four of those 6-inch squares could fit into a square foot. In this case, the plotted points should be standardized in what we call a u control chart, described later in this section.

To describe this kind of count data, we usually rely on the Poisson distribution, which is pictured in Figure 5.11. This distribution is appropriate to use when we are counting the frequency of rare occurrences (defects in the product) that have a nearly uncountable number of opportunities to happen. For this reason, the Poisson distribution is sometimes called the law of small numbers. In Figure 5.11, the average number of defects found in sample units is 2.5; this number is chosen so we can more easily compare this distribution of defects from a c chart to those found in an np chart where the sample size is 10 and the nonconforming fraction is 25% (so 2.5 defects are expected in both samples). The distribution of defects from the c chart (based on the Poisson distribution) will have larger variance than those from the np chart; in this case the standard deviation is 15% larger for the c chart.

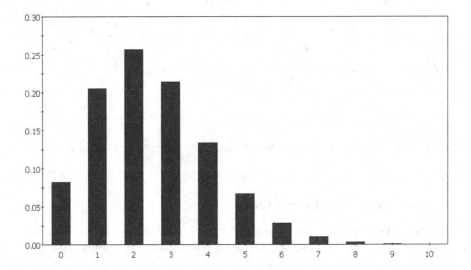

**Figure 5.11**   Distribution of the number of defects found in a unit when the average number of defects per unit is 2.5.

To make a c chart, we need to sample units of the same size at each time point. For the fabric swatches, all of the sample units are five meters in length, and the c chart is appropriate in such a case. For these applications where we count the number of defects in samples that differ in size, we need to distinguish homogenous units in the sample (for example, 1-meter lengths of fabric) and divide the number of defects by the number of units in the sample. When controlling the average number of nonconformities per inspection unit is of interest, the u control chart is the appropriate choice. For the u chart, we record the number of defects along with the number of homogenous units, but only the average is plotted.

## Example 5-6: Fabric Inspection

A fabric manufacturer wishes to control the number of flaws in 46" x 54" swatches of fabric that will be transformed into curtains. Twenty swatches are inspected, and the number of flaws found in each curtain is listed in the file `Fabric.ezs`. Using the same instruction as in the previous examples, the c control chart is given below. The process, plotted in Figure 5.12, is clearly in control with respect to the $3\sigma$-limit imposed on the UCL.

• • •

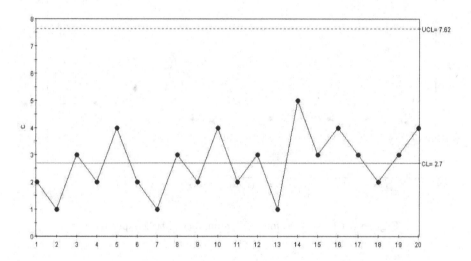

**Figure 5.12**    c chart for fabric defect data.

## Example 5-7: Delivery Trucks

A delivery company wishes to assert more control on the average number of speeding tickets acquired by their delivery trucks. The company recorded the total number of speeding tickets and the number of delivery trucks operated on each day for 20 consecutive work days. The data are given in `Tickets.ezs`. This example is different from the last because a different number of units are inspected each day, and the sample size (number of trucks operating) must be recognized when computing the average number of tickets per truck, so a u chart is appropriate. The chart in Figure 5.13 is based on average number of tickets per driver, and because the number of trucks operating each day changes, our uncertainty changes correspondingly, so the control limits are not constant. The u control chart in Figure 5.13 shows that the process is in control.

• • •

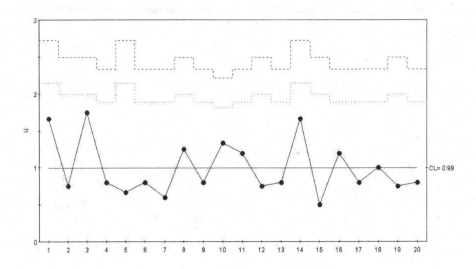

**Figure 5.13**   u chart for average number of tickets per truck.

## 5.5   DECIDING WHICH CHART TO USE

We have reached the end of the chapter and have a chance to look back and wonder how we can put all of this knowledge together. We have looked over eight different statistics to chart for our process, and our choice of control chart depends on the problem at hand. There are two types of data (variable data, attribute data) that are easy to distinguish, so this simplifies the problem somewhat. For variable data, we have three charts we discussed at length: the $\bar{x} - R$ chart, the $\bar{x} - s$ chart, and the x-MR chart. Our choice of these three depends on how much data we have available to analyze in the process. If we see only one observation at a time, we must choose the x-MR chart. If we have multiple measurements at each sampling point, but less than ten, we should use the $\bar{x} - R$ chart. This is probably the most common scenario we see in practice, so we featured it as our first control chart. If we have 10 or more observations at each sampling point, it is better to use the $\bar{x} - s$ chart because it will give us a more precise analysis.

In the case we have attribute data (or count data), there are four charts that might be useful: u chart, c chart, p chart and npchart. It's a little trickier deciding which of the four is the right one to use, but it has to do with what we are counting. For example if we are counting items or units, we will use a p chart in the case in which the sample size can vary at different sampling times. If the sample size is always the same then it is more fair to compare the counts at each sampling time, and the np chart is the right one to use.

The c chart and u chart are for counting *occurrences*. This differs from the other attribute charts where there is always a set number of things that could be counted at every sampling point. For example, if we have 100 golf club shafts, we can count the number of defective clubs found to be between zero and 100. That is why the p chart of np chart was used for that problem. On the other hand, if we are counting the number of flaws found on a swatch of fabric, we have no fixed upperbound, and we use the c chart for the case in which all the items are identical in size or the u chart if some items are bigger than others (and so have a greater propensity to have flaws).

Figure 5.14 offers a graphical summary of this decision process, based on a decision tree that can help you select the right chart for the right process data.

## 5.6 WHAT DID WE LEARN?

- Depending on the kind of data is generated by the process, there are seven different control charts that can be used (see Figure 5.14).

- A control chart is based on a centerline, representing the target value, along with lower and upper control limits that characterize when the process is in control.

- The standard $3\sigma$ control chart allows 0.27% of the items to go outside of the control limits when the process is known to be stable.

- Warning limits are used to notify the process manager when outputs fall two standard deviations away from the target value.

- Average run length is the number of times a stable process will be considered in control before a false alarm occurs.

- To monitor the mean value of the process output, a $\bar{x}$ chart is used.

- To monitor process variability, an s chart is used unless the sample sizes are small (less than 10), in which case an R chart is used.

- If only one measurement is allowed at sampling points, the $\bar{x}$, R and s charts cannot be effectively used, but a moving-range (MR) control chart can be constructed based on the differences of successive measurements.

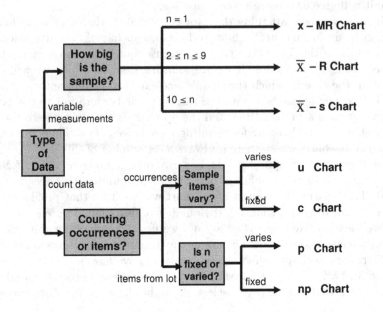

**Figure 5.14** A decision tree for summarizing how you choose the appropriate control chart for the given data in the SPC problem.

- To chart a process in which output items are inspected on a pass/fail basis, we can use an attribute control chart such as the p control chart or np control chart.

- For monitoring the number of defects in the process output, the c control chart should be used when the defect count is not limited on any particular item.

## 5.7   TEST YOUR KNOWLEDGE

1. For an in control process, roughly what proportion of the observed output will fall outside of the warning limits of the standard Shewhart control chart?

   a. 0.3%

   b. 1.0%

   c. 2.0%

   d. 5.0%

**2.** What is the average run length ($ARL$) of an in control system that has a false alarm rate of 0.000819?

   a. 74

   b. 185

   c. 370

   d. 1221

**3.** For the tissue paper production in Example 5-1, which of the following potential output measurements represent attribute data?

   a. machine direction tensile strength (in grams/76mm-2plies)

   b. weight of a roll of tissue (in grams)

   c. inspector's softness rating (very course, course, soft, very soft)

   d. conformity status (passed or not) based on various measured qualities

**4.** On a silicon wafer, 30 chips are constructed and the defective chips are detected and removed before further processing occurs. To graph the control of this process, we should use

   a. Shewhart control chart

   b. np control chart

   c. p control chart

   d. c control chart

**5.** A window manufacturer makes glass panes in three sizes: 2' x 1', 2' x 4', and 3' x 4'. Each pane is inspected for minor flaws, and the number of defects found on each inspected pane is recorded. How should we chart the control of this process?

   a. Shewhart control chart

   b. u control chart

   c. p control chart

   d. c control chart

**6.** At particular time points during a production process, outputs are sampled and measured to see if they are conforming to specification. In general, how many units are sampled at each time point?

   a. one or two

   b. four or five

   c. ten to twenty

   d. thirty

**7.** To achieve an average run length ($ARL_0$) of 200, what is the maximum percentage of false alarms allowed by the process?

   a. 0.50 %

   b. 1.00 %

   c. 2.00 %

   d. 5.00 %

**8.** Five items per batch are inspected on a pass/fail basis. If the nonconforming fraction of the items is 10%, what is the approximate probability that any particular batch of items produces five conforming products?

   a. 0.01

   b. 0.13

   c. 0.24

   d. 0.60

**9.** In Example 5-6, a c control chart is generated for fabric defect data. According to the eZ SPC output, an out of control signal would be generated if seven or more successive points appeared on one side of the centerline. What were the most successive points on one side of the line for this example?

   a. 2

   b. 3

   c. 4

   d. 5

**10.** In example 5-6, if you produce descriptive statistics (under the Analysis tab) in eZ SPC, you will find the mean number of defects in 20 samples is 2.7. According to the c control chart, the distribution of defects should be Poisson. One important characteristic of Poisson distributed data is that the mean and the variance should be close to the same. In this case, the sample variance is

   a. less than half the value of the mean

   b. less than 90% of the mean value

   c. exactly the same as the mean

   d. more than 50% larger than the mean value

# EXERCISES

**5.1**   The wafer size for a semiconductor has gradually increased to improve throughput and reduce cost with the current state-of-the-art fabrication (or fab). In 2010, the global leading semiconductor fabrication plants use 300 mm diameter (11.8 inch, usually referred to as "12 inch" or "Pizza size") wafers for production. The wafer thickness, which is measured at the center of the wafer with a wafer thickness gauge, is also an important parameter in this process because a very thin wafer is prone to warpage or breakage while a very thick wafer may not allow processing in all wafer fab equipment. The thickness of standard 300 mm wafer is 775 m. A wafer manufacturing company monitors and controls the thickness of wafer by $\bar{x}$-R control charts, and the data are given in Wafers.ezs.

A quality engineer took 25 samples of sample size 5 for analyzing the process. Select data and click

Graph ▷ $\bar{x}$-R control chart

Select For Analysis in the window that appears, then click Select (order) default and click OK. The chart should show that the process is in control.

**5.2**   We are going to work with the same data as the first exercise (Wafers.ezs). Again, with all data, select Graph ▷ $\bar{x}$-R control chart. Rather than selecting For Analysis, instead choose

For Control

and type in the UCL, CL, LCL for the $\bar{x}$ control chart and R control chart using values from the previous exercise. Is the process still in control? For this kind of manufacturing problem, new control levels are computed because periodically we need to reset the UCL, CL, and LCL to adjust control charts to the current process because the process variability might be reduced.

**5.3**   A paper manufacturing company produces a standardized copy paper with thickness of 1.0 mm. Uniformity of paper thickness is extremely important to the company, and they regularly check the copy paper thickness during the production process. Data from their production process is given in CopyPaper.ezs. Construct an $\bar{x}$-R chart for this data and determine if the process is in control. Remember, you have a fixed and known target value in this example.

**5.4**   A water bottling company produces bottled water with a certain amount of fluoride, which is important for good oral health. Fluoride can occur naturally in source waters used for bottling or it can be added during the filtration process. The bottling company controls the amount of fluoride in the individual bottles so that the fluoride level is greater than 0.6 milligrams per liter (mg/L) but no more than 1.0 mg/L. Twenty bottles are randomly sampled from the plant and measurements of fluoride on bottled water are given in the file `Fluoride.ezs`. Apply the x-MR control chart to this process and determine if it is in control with respect to fluoride levels.

**5.5** Earphones are generally inexpensive and are favored by audio listeners for their portability and convenience. The manufacturer of low-cost earphones randomly selected 25 samples of their earphones for inspection and counted non-conformities. The data are given below:

1 2 2 1 2 3 1 1 0 0 1 0 0 2 1 0 0 1 0 0

Construct an appropriate control chart for this process.

**5.6** A paper box manufacturing company produces flat rate boxes for the US Postal Service. The company packs 100 boxes into one retail unit. The printing quality of a box is checked regularly, and in a recent inspection, twenty units were sampled, and the number of defective boxes per unit is recorded in the file `Boxes.ezs`. Construct a p control chart for the process and comment on the results.

**5.7** A wholesale store seeks to minimize the number of bounced checks received from their customers for payment. The number of checks paid per day and the number of bounce check are recorded for thirty working days, and the data are listed in the file `Checks.ezs`. Construct an appropriate control chart for this process.

# CHAPTER 6

# ADVANCED CONTROL CHARTS

The $\bar{x}$-R, $\bar{x}$-s control charts are known to be effective in detecting any dramatic change that occurs in the process. Consider another potential problem with a process: what if the change that occurs is steady, persistent, but not as obvious in terms of the amount of change in the process from one time point to the next. For example, due to a variety of possible assignable causes, a manufacturing process can slowly degrade, causing slightly more nonconforming products in each time period. Compared to the large shifts in the process that are detected with the previously discussed control charts, these small or moderate shifts in the process output can only be detected if we accumulate information from a series of past outputs when we make a decision whether the process is in control. In order to resolve the unresponsiveness of Shewhart control charts in terms of small shifts in the process, the cumulative sum (CUSUM) control charts and the exponentially weighted moving average (EWMA) control charts are very good alternatives. The CUSUM and the EWMA control charts are typically used with individual observations.

*Basic Statistical Tools for Improving Quality.* By Chang W. Kang and Paul H. Kvam **125**
Copyright © 2011 John Wiley & Sons, Inc.

## 6.1   CUSUM CONTROL CHART

CUSUM (pronounced *Q-sum*) is short for cumulative sum, which means the values plotted on the control chart are accumulated and averaged as the process continues to produce outputs that are sampled in successive stages. In accumulating past data, each past sample has equal weight in how the control chart is constructed. The CUSUM chart is made to be sensitive to slow, insidious changes occurring in the process, unlike the Shewhart charts. For this benefit, the chart sacrifices some ability to detect a dramatic change that occurs in the process, especially if the process has been operating for a long time. While the Shewhart charts can reveal such a change in the process with a spike in the control chart, a detectable change in the plotted average can be more muted in the CUSUM chart because most of the weight in the charted value comes from past samples, when the process was presumably in control.

The CUSUM control charts use the equally weighted information in the sequence of samples to monitor the process. At each stage, we observe how the process deviates from the target value. The CUSUM control chart uses the cumulative sum of these deviations. This enables the process manager to effectively detect a gradual shift in the process, even if it is small in magnitude.

The CUSUM control chart is typically constructed for individual observations. When we construct the tabular CUSUM control chart, two statistics $C+$ and $C-$ are calculated:

$C+$ = Accumulation of deviations upward from the target value
$C-$ = Accumulation of deviations downward from target value.

As the process manager, we can choose values for two important parameters, $K$ and $H$, allowing us to construct the tabular CUSUM control charts in a way that is most beneficial to us. $K$ is called the *reference value* and H is called the *decision interval*. We need to choose the reference value $K$ to be one-half of the shift magnitude that we want the control chart to detect. In general, the shift is expressed in standard deviations and the CUSUM control chart is effective when the magnitude of shift is $\sigma$ (one standard deviation). The value for $K$ is 0.5 times the process standard deviation. As a rule-of-thumb, a reasonable value for $H$ is 4 or 5 times the process standard deviation.

### Example 2-1: LCD Panel Quality Data

In the LCD Panel Quality Data, two columns of data (for Company A and B, respectively), represented a quality measurement of the distance between marks of tape for a tape carrier package. Specification limits were set to

$25.22 \pm 0.01$ $mm$. So if we want to detect a mean shift of 0.01 $mm$ in this process, we set $K$ to half of this value, 0.005. If the specification limit was not already set, we could attempt to assign $K$ to be half the estimated standard deviation ($s$) of the process. In this example, we can find $s$ using eZ SPC by choosing `Descriptive Statistics` under the Analysis tab. There is little variability in this stable process, where $s = 0.0034$, and we would alternatively assign $K = 0.0017$.

We can set the control limits for the CUSUM chart by changing the value of the decision interval ($H$). If the CUSUM statistics plot outside the decision interval $H$, then the process is considered to be out of control and we would then trace back to find the first non-zero CUSUM statistic in the sequence of samples. The process shift originated at that sample.

To illustrate the tabular CUSUM, consider the 12 observations listed on the left side of Figure 6.1. These are process output measurements from a hypothetical process that has a target mean of 100 units, with a known standard deviation of 10 units. Suppose we want to detect any shift in the mean that is more than 20 units from 100. $K$ is equal to half of this shift (10), divided by the standard deviation (10): $K = $ (half shift)$/\sigma = 10/10 = 1.0$. This reference value defines a "comfort zone" for output measurements defined by the target mean, plus or minus $K$. In this example, our comfort zone is (99, 101). If we assign the decision interval to be 4, then $H = 4\sigma = 40$. When we accumulate deviations in $C+$ and $C-$, the out-of-control signal is implemented if either value exceeds 40.

| Data | 90 | 95 | 97 | 99 | 100 | 101 | 103 | 105 | C- | C+ |
|------|----|----|----|----|-----|-----|-----|-----|----|----|
| 98 (-1) | | | | 1 | | | | | 1 | 0 |
| 105 (+4) | | | | | | | | 4 | 0 | 4 |
| 103 (+2) | | | | | | | 2 | | 0 | 6 |
| 98 (-1) | | | | 1 | | | | | 1 | 5 |
| 97 (-2) | | | 2 | | | | | | 3 | 3 |
| 90 (-9) | | 9 | | | | | | | 12 | 0 |
| 85 (-14) | 13 | | | | | | | | 25 | 0 |
| 89 (-10) | 10 | | | | | | | | 35 | 0 |
| 99 (0). | | | | | | | | | 35 | 0 |
| 92 (-7) | | 7 | | | | | | | 42 | 0 |
| 94 (-5) | | 5 | | | | | | | 47 | 0 |
| 95 (-4) | | 4 | | | | | | | 51 | 0 |

**Figure 6.1** CUSUM Table to illustrate computations.

In Figure 6.1, the first value (98) is two units under the target value, and one unit under the comfort zone. In the column for $C-$, we enter 1 because the observation represents a downward deviation from the target value, and it is one unit under the comfort zone. We do not subtract anything from $C+$ because the accumulated deviations cannot be below zero. The next observation (105) is four units over the comfort zone, so $C+$ goes from zero to 4. For $C-$, we subtract four from the previous value (1), and because this is less than zero, we set $C- = 0$. The next observation (103) adds 2 to $C+$, and the following data (98) adds 1 to $C-$ while subtracting 1 from $C+$. In this case, $C+$ is still positive, but reduced from 6 to 5. After this, we see a trend of values that are less than the target value, so $C-$ keeps accumulating these negative deviations. After the 10th observation (92), $C-$ crosses the threshold of 40 ($C- = 42$), so the out-of-control signal is executed. The process manager has been alerted in the 10th stage, but the proper action is to go back to where this streak of $C-$ started (at the fourth stage) and search for an assignable cause to this trend. Once corrective actions are finished, we restart the CUSUM back to zero and continue sampling.

$\bullet\ \bullet\ \bullet$

### Example 6-1: Sanitary Pads

A company that manufactures feminine hygiene products makes a protective pad that depends on high-seam seal strength as a measure of quality. The target value for pad seal strength is 400g/2 inches-width and it has only a lower specification limit of 300g/2 inches-width. Seal strength is affected by the temperature and pressure of the press roll and retention time. To ensure any process degradation can be efficiently detected, the quality engineer on the assembly line wishes to control the end seal strength by employing a CUSUM control chart. The data from the process are given in `Pad.ezs` from your applications folder. In this example, the out-of-control signal occurs at the eighth time frame, but because no corrective action was taken, the chart was not reset to zero. To find an assignable cause for the process being out-of-control, the process manager will go back to the second time frame, when the positive deviations first started to accumulate.

$\bullet\ \bullet\ \bullet$

What we are calling the CUSUM is technically defined as *tabular CUSUM*. There are actually two types of charts for representing cumulative sums, the other one being the V-mask CUSUM. However, the tabular CUSUM is more

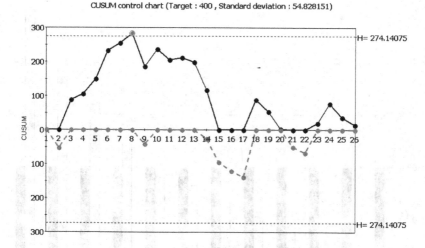

**Figure 6.2**   CUSUM diagram for pad seal strength.

preferable. For that reason, we focus solely on the tabular CUSUM procedure in this book.

We can see the CUSUM is not as simple as the Shewhart charts, so it's not as easy to chart or to interpret. But the CUSUM has shown to be better at defecting small shifts in the process mean, and that is why it has become one of the most frequently used control charts for manufacturing and other industries.

## 6.2   EWMA CONTROL CHART

Like the CUSUM control chart, the *exponentially weighted moving average* (EWMA) control chart uses an average of past measurements (usually individual observations) from the process in order to signal whether the process is in control or not. Unlike the CUSUM, the EWMA gives less weight or importance to values as they go back in time, so the plotted value on an EWMA control chart is actually a *weighted* average. In this way, the EWMA serves as a compromise between the Shewhart charts that give no weight to past measurements in the chart analysis, and the CUSUM, which treats every past measurement equally. For example, Figure 6.3 shows the relative weight one such EWMA chart gives to the past measurements after 12 measuring stages have been completed. The last observation (getting the most weight) represents the most recent measurement, the next-to-last one is the next most recent past measurement, and the initial bar on the left represents the first measurement made in the charting procedure.

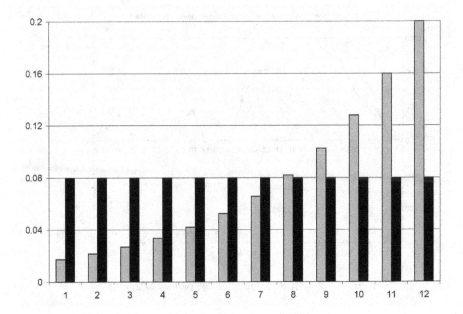

**Figure 6.3**    Relative weighting for EWMA (gray) compared to CUSUM (black).

The amount of weight given to past data is determined by a weighting factor that we will designate with the Greek letter $\lambda$. We choose the weighting factor to be a number between zero and one, and represents how the weights in the figure are assigned. For example, in Figure 6.3, $\lambda = 0.20$, which means each measurement receives $(1 - \lambda) = 0.80$ or 80% of the weight in the plotted average compared to the one that follows it next in time. This recursive way of assigning weights leads to a geometric (or exponential) progression, which can be seen in Figure 6.3.

The choice of $\lambda$ can greatly affect the performance of the EWMA control chart. In Figure 6.4, two different weighting schemes are plotted. With the smaller value of $\lambda = 0.2$ (black bars), the weight is distributed more evenly than it would be with $\lambda = 0.4$ (gray bar). This is how the process manager can control the EWMA chart, in terms of how much importance is given to past measurements in the process. For example, if the process will not retain any residual effects past three or four time points, it would be better to assign a relatively larger value of $\lambda$ (0.5 or 0.6) so that only the past three or four measurements receive any significant weight in the final output measurement.

Because the EWMA control chart relies on weighted averages of all past and current samples to monitor the process, it can be very effective in detecting a small shift in process output. It also has a quality of *robustness*. While most charts require the assumption that the distribution of sample measurements is approximately bell-shaped (or normal), the EWMA seems to perform well

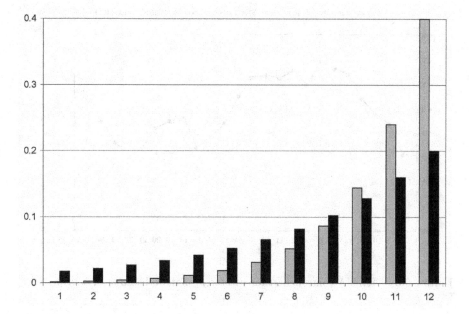

**Figure 6.4**   EWMA weights using two different weight constants: $\lambda = 0.2$ (black) and $\lambda = 0.4$ (gray).

even when that is not the case. We discussed hypothesis tests for the normality of the data in Section 7 of Chapter 3. In general, $\lambda = 0.05$, $\lambda = 0.1$, and $\lambda = 0.2$ are popular choices that allow information from past samples going back eight or nine time periods to be included in the weighted average.

Along with the weight factor $\lambda$, we can choose a second factor (denoted L) that directly affects the control limits. The choice for L is based on $ARL_0$ (the average run length for a process that is in control). If we stay with $ARL_0$ values consistent with those we used in constructing Shewhart control charts, the rule-of-thumb choice for L is between 2.6 and 3.0.

**Example 6-1, continued**

Figure 6.5 shows the EWMA control chart on the pad strength data that was previously analyzed using the CUSUM procedure. In this case, by emphasizing the most recent observations, the process was never considered to be out of control.

• • •

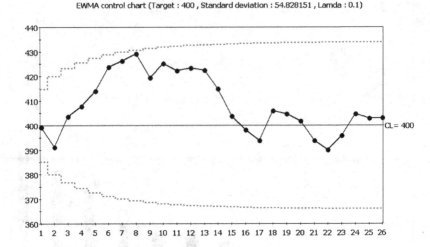

**Figure 6.5**   EWMA control chart for pad strength data, based on $\lambda = 0.10$.

## CUSUM versus EWMA

There are numerous studies that compare the effectiveness of the CUSUM and EWMA charts. Depending on the type of application, they both have advantages and disadvantages, so no single chart can be recommended for all process problems in which we are trying to detect a small shift in the process mean.

For specific applications, the EWMA can outshine the CUSUM if the exact right weighting factor ($\lambda$) is selected. The problem, of course, is that we never know exactly what the right weighting factor should be. The EWMA is robust because it works well even if the distribution of the data is not normal.

The CUSUM has advantages, too. It's generally simpler and easier to interpret than the EWMA. In general, it will be easier for the public to believe the conclusions of a chart if it can be more easily explained. For this reason, the CUSUM has been more widely used in practice. Remember, both the CSUSUM and EWMA control charts are typically used with individual observations.

## 6.3   CV CONTROL CHART

When the process is stable, the process data exhibit a constant mean and variance. We use the $\bar{x}$ control chart for monitoring the process mean and the R or s control chart for monitoring the variation of the process. To monitor these two characteristics simultaneously, we generally assume that the variance in the process output is not a function of the mean. In other words, if the mean drifts from the target, the previous control charts treat the process variance as if it is unaffected by this drift. In some circumstances, this is not realistic. In some manufacturing processes, the variance can change, depending on the process mean. For example, if the output value reflects process yield, an increased mean might naturally lead to an increase in variance.

When the variance is a function of the mean, the coefficient of variation is an appropriate measure for process variability. In Chapter 3, we introduced the sample coefficient of variation

$$CV = \frac{s}{\bar{x}}$$

as a useful measure of variability for processes in which the output mean changes from one setting to the next. Just as the standard deviation can be misleading in such a setting, the $\bar{x} - R$ charts or the $\bar{x} - s$ charts are not appropriate for monitoring the process mean and variation. Instead, we use the coefficient of variation (CV) control chart, which was first introduced by Kang et al. (2007).

As you might expect, the CV chart plots the CV statistic at each time plot, and we monitor the process to make sure the coefficient of variation remains constant throughout the process. The CV chart is commonly used in environmental monitoring, where pollution or effluent release is monitored but cannot be controlled, and the mean amount can change dramatically due to seasonal effects or changing regulations.

The CV chart is a "specialty chart" that should be applied when simpler charts are not appropriate. The CV chart keeps track of the process variability and possible shifts from the process mean at the same time. If these two process characteristics do not interact, it would be better to monitor them individually, perhaps using a $\bar{x} - R$ chart or $\bar{x} - s$ chart.

## Example 6-2: Effluent Monitoring

The U.S. Environmental Protection Agency is involved with effluent monitoring at power plants and national research laboratories. Suppose we are monitoring airborne radioactive contaminants at a plant that experiences great fluctuation in waste management activity. At specified time intervals, five air quality measurements are sampled around the plant Effluence changes according to the amount of processing done at the plant, which is not constant day to day. The data are listed in CV.ezs.

The $\bar{x}$ chart is not applicable for detecting a shift in mean discharge due to this expected fluctuation. The $\bar{x}$ chart in Figure 6.6 makes this point clear; the average emission changes dramatically between time points, so the $\bar{x}$ chart,

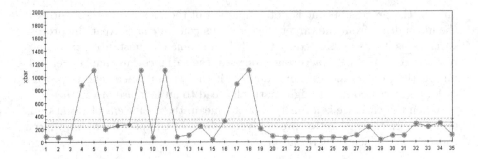

**Figure 6.6**   Chart of average contaminant emission level using the $\bar{x}$-chart, which is not sufficient in monitoring this process.

R chart and s chart fail to tell us what we need to know about the process. Instead, we can take this output and standardize it with the CV control chart. To plot the CV control chart, open the data file, select the data and go to

```
coefficient of variation (CV) control chart
```

under the Graph tab. The control chart in Figure 6.7 is generated by using the default values in the eZ SPC procedure. The output will list the computed CV values along with the upper and lower control limits for the process. In this example, the contaminant level changes across the sampling times, but the CV statistic is contained within the control limits.

• • •

## 6.4   NONPARAMETRIC CONTROL CHARTS

There are numerous kinds of advanced control charts we are not going to cover in this book because our purpose is to focus on the fundamentals for quality improvement. There is one more category of control charts, however, that we feel obliged to mention: *nonparametric control charts*. A nonparametric statistical procedure is one that does not presume the data are from the normal distribution. These are robust procedures which are considered safe to use since they work in every kind of situation.

We saw that for detecting a gradual shift in the process mean, the EWMA chart is typically robust. But for detecting change points where the process mean changes abruptly, Shewhart charts are more appropriate. Although any Shewhart control chart is effective if the data are from a normal distribution,

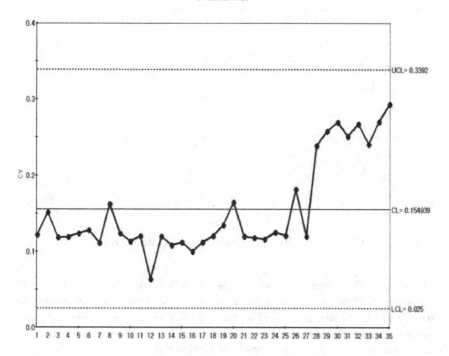

**Figure 6.7**    CV control chart for contaminant emission level in Example 6-2.

the results of an X-MR chart can be misleading if the data a highly skewed. For such sets of data, the program manager needs a nonparametric chart.

Although eZ SPC does not come equipped with an array of advanced control charts, it is easy to contrive a simple nonparametric chart to replace the $\bar{x} - R$ chart in case the data are highly skewed. We will describe one such chart for sample medians here. Recall that the sample median refers to the sample value that ranks in the middle. If there are 11 observations in the sample, the median ranks as 6th highest (and 6th lowest) out of 11. If the data are from a skewed distribution, the sample mean and the sample median can vary (even though they both are constructed to gage central tendency).

## Example 6-2: Effluent Monitoring

Figure 6.8 shows a histogram of the effluent data, which is highly skewed. There is no way you can look at that histogram and see a mound shaped

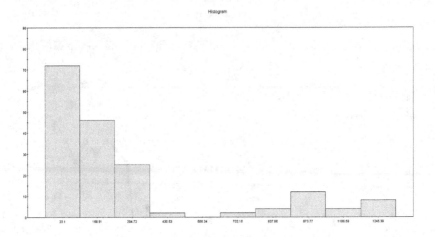

**Figure 6.8** Histogram for effluent monitoring data.

distribution. The sample median in this example (116) is less than half of the sample mean (289), which is influenced by the outlying measurements that exceed 800. As we saw in the last section, the CV chart was more appropriate than a standard $\bar{x} - R$ chart. But we can also illustrate how a nonparametric chart such as the median chart works with this data.

Rather than graphing the sample averages (based on the five subgroups), we can alternatively order the five observations at each of the 35 stages and plot the medians. In the first measurement period, we obtained the data (68.1, 90.9, 91.2, 84.1, 93.8). The middle value is 90.9, which is larger than the sample mean of 85.6. The range ($R = 25.7$) is used in the median chart just like it is used in the $\bar{x} - R$ chart.

To set up the control limits for a (nonparametric) medians chart, we compute the average of all 35 sample medians ($\bar{M} = 290.866$) and set up the control limits according to how large the sample sizes are at each stage:

$$\bar{M} - T_n \bar{R}, \ \bar{M} + T_n \bar{R}$$

where $\bar{R}$ is the average of the 35 sample ranges, which is $\bar{R}$=96.317. The constant $T_n$ depends on how many ($n$) observations are measured at each stage:

| $n$ | 3 | 5 | 7 | 9 |
|---|---|---|---|---|
| $T_n$ | 1.187 | 0.691 | 0.508 | 0.412 |

With $n$=5 observations, we use $T_n = 0.691$. The center line for the median chart is set at $\bar{M} = 290.866$, and the lower and upper control limits are $(290.866 - 0.691 \cdot 96.317, 290.866 + 0.691 \cdot 96.317)$, or $(224.31, 357.42)$.

In this case, the median control chart is similar to the $\bar{x}$ chart, even though the means and medians are not the same for the 35 samples. But compared to the control limits graphed in Figure 6.6, the control limits for the median control chart are more than two times wider. There is a good reason for the disagreement between the $\bar{x}$ chart and the median chart. The $\bar{x}$ chart offers narrower control limits because it assumes the data have a normal distribution. This is an incorrect assumption, and this "false confidence" leads to narrower intervals on the chart. Figure 6.9 shows the normal probability plot for the 35 sample means. Due the highly non-linear trend in the plotted points, the plot strongly suggests that the data are not normally distributed.

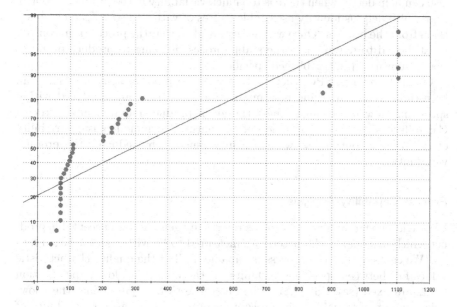

**Figure 6.9**    Probability plot for effluent data based on sample medians shows data are not from a normal distribution.

It's important to add at this point that the data are best explained by the CV plot, and not the median plot or the $\bar{x}$ plot. This is because the coefficient of variation $(CV = \sigma/\mu)$ is more pertinent to this kind of data in which variability and central tendency are so intertwined. While the nonparametric (median) chart is preferred to the $\bar{x}$ chart in this situation, the CV chart is still the best one to use.

$\bullet \quad \bullet \quad \bullet$

Other nonparametric control charts are based on using the ranks of the data instead of the actual measurements. While researchers agree that many of the advanced process problems should be addressed using nonparametric charts, there is little evidence that program managers actually use them in practice. This will probably change over the course of years, but the change might be slow.

## 6.5 PROCESS CAPABILITY

Up to this point, we have used statistical process control to monitor a process in order to make sure it produces acceptable products continuously. Control charts not only allow us to detect departures from the target mean, but they also can help detect when there is too much variability in the product. Control charts also help us identify assignable causes of variation so we can eliminate them from the process. Once we have established that the process is in control, we still need to figure out if the stable process is going to produce products that conform to pre-established specifications.

Our next goal in SPC, then, is to assess the *process capability*. This can be estimated by observing how many process outputs are produced within specs and how many fail to meet the given product requirements. Process Capability is generally a function of the natural (common-cause) variability of the product output, so we can achieve success by reducing the product variability.

### Process Capability Analysis

We are interested in how the process repeatedly generates an acceptable product which has quality characteristics measured to be within specification limits. We can say that the process is capable if all of the quality characteristic values fall between specification limits. If the upper and lower specification limits are fixed (say USL, LSL), the difference USL − LSL defines the allowable amount of variability in process outputs. If the natural variability of a process output is characterized with standard deviation $\sigma$, we can recall from Chapter 4 (Shewhart Control Charts) that a stable process will generate products within (plus-or-minus) $3\sigma$ of its target mean 99.73% of the time. To measure process capability, we can simply contrast the allowable tolerance (USL − LSL) with the $6\sigma$ interval that contains 99.73% of product outputs.

A statistical measure called the process capability index, $C_p$, is defined as the ratio of the difference of these two values:

$$C_p = \frac{USL - LSL}{6\sigma}.$$

If the standard deviation is unknown, we use the computed process standard deviation $s$ as a substitute. Because the difference between the upper and lower specification limits is fixed, if the process standard deviation decreases, then the process capability index becomes larger. In other words, the larger the $C_p$ value, the less variability is exhibited in the process output. Once the process is in control, the process manager can concentrate on improving the process by increasing $C_p$. Depending on the kind of situation of the actual process, the recommended (minimum) process capability can change to 1.0 to 2.0, in general. Montgomery (2008) recommends different process capability requirements for different situations, from 1.33 for existing processes up to 2.00 for six sigma quality processes. See Figure 6.10.

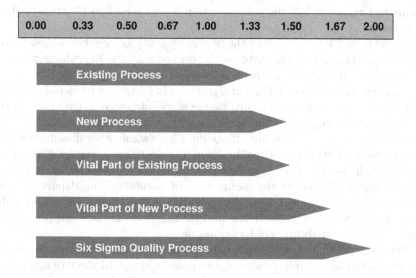

**Figure 6.10**  Process capability requirements for different situations.

If $C_p$ is significantly less than one, too many defects will be produced and we would no longer believe the process was stable. To conduct a process capability study, the following assumptions are needed:

**Assumption 1.**  The quality characteristic follows the bell-shaped curve of the normal distribution.

**Assumption 2.**  The process is in statistical control.

**Assumption 3.**  The process mean is centered at the mid point of two sided specification limits.

Most likely, the second and third assumptions have already been addressed. To find out if the quality characteristic has a distribution that is bell-shaped like the normal distribution, we could plot the histogram of the data (see Figures 3-4 and 3-5 in Chapter 3, for example). A better option is to use a *normal probability plot*, which is a simple graphical procedure invented for the sole purpose of checking the normal distribution assumption. We first considered the probability plot Chapter 4. To create a normal probability plot of a column of data in eZ SPC, simply choose the

`Normality Test`

procedure under the Graph tab. The procedure window allows you to specify mean and variance, but the default action is to have eZ SPC estimate these parameters.

Recall from Chapter 4.4 that the normality test produces two pages: `Chart` and `Data`. The data page displays the outcome of the hypothesis test $H_0$: the data are distributed normally versus $H_1$: distribution of the data is not normal. See the section on *Tests of Hypothesis* in Chapter 4 for details on these statistical tests. From this output, we are more interested in the chart. If the process is producing data with a bell-shaped distribution, the probability plot will approximate a straight line. If the data have some other distribution, the plotted points will curve noticeably. Even if the data are normally distributed, the line will never really be perfectly straight, so your discretion is part of the judgment. To reject the assumption of normality, the departure from straightness must be clear if not dramatic. If the process data is not normally distributed, we must terminate the study at this point because our predictions for the process capability will be erroneous.

Since the process capability should represent the real ability of the process for producing acceptable products, the process standard deviation must be calculated when the process is in statistical control. In order to check the stability of the process, the most common way is to use the $\bar{x} - R$ control chart.

## Example 5-1: Tissue Paper Strength.

In Example 5-1, we monitored the quality of tissue paper by measuring tensile strength (MDTS). The target value was 937 grams/76mm-2plies, and by using a $\bar{x} - R$ chart, we decided the process was in control. Suppose the product specification limits are [825, 1060]. We can obtain a process capability analysis in eZ SPC by selecting

`Process Capability Assessment`

under the Tools tab. Enter the upper specification and lower specification and click OK. eZ SPC produces the chart in Figure 6.11. From the data output, we estimate the standard deviation of the in-control process to be 10.6485 grams/76mm-2plies with $C_p = 3.858$. The value of $1/C_p$ is 0.2592, so the distribution of MDTS measurements falls within the tolerance limits, taking up only about one fourth of the band of values allowed by the tolerance limits.

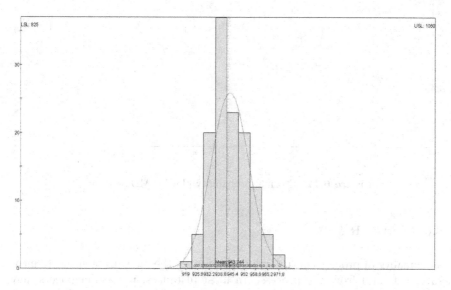

Cp : 3.857972  Cpk : 3.817127  Pp : 3.678152  Ppk : 3.639211

**Figure 6.11**   Process Capability Analysis Result for MDTS data.

In Figure 6.12, a normal probability plot for the MDTS data reveals no strong non-linear trend in the graphed data. Without obvious curvature, we have no strong reason to doubt that the measurements follow a normal distribution.

• • •

Along with the $C_p$ statistic, the normal probability plot is an essential part of process capability analysis. We will return to this probability plot in the next chapter, where it will be used to validate the application of a regression procedure.

Even though $C_p$ is a good measure of process capability, it does not account for the process that is off-center. If the process mean is shifted away from the target value but the process standard deviation is unchanged, $C_p$ value is same. $C_{pk}$ value must be considered in this case. $C_{pk}$ accounts for the process that is off-center. Therefore process engineers always check $C_p$ and $C_{pk}$ at the same time.

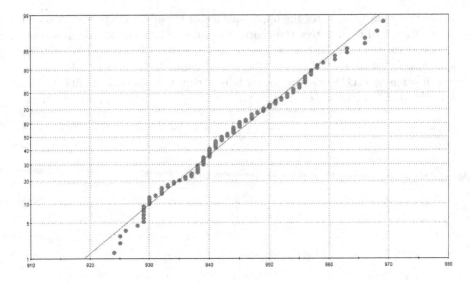

**Figure 6.12**   Normal probability plot for MDTS data.

## 6.6   GAGE R & R

The quality of process control will depend upon the analysis of accurate measurement data from the process. In most instances, process engineers may assume the data collected for the process control are accurate. This means that the process engineers understand that there is no variation in the gages that measure the process. But most gages are subject to variability due to small variations of the gage itself. Also, measurement differences can be due to appraiser variability. The analysis of measurement data obtained by an unacceptable gage or by an appraisers' misuse may lead to inaccurate conclusions about the state of the process. In order to reflect the process variation only in the control data, we must try to minimize gage variation when we detect it. Therefore, it is important to analyze the gage and appraisers periodically.

There are four primary characteristics for the evaluation of gages:

1. **Bias**: Bias describes the difference between the observed average of the measurements and the reference value. Sometimes biased measuring devices can be shifted, or calibrated, to minimize the difference between the reference value and the measurement average.

2. **Repeatability**: Repeatability is the measure of variation in readings of the same part by the same appraiser. If the gage has poor repeatability, then there are internal gage problems. It is also known as equipment variation.

3. **Reproducibility**: Reproducibility is the measure of variation in average measurements when the gage is used by different people. It is also known as operator variation and represents the variation due to the measuring system. A large value of reproducibility might indicate lack of training in the appraisers.

4. **Stability**: Stability is the measure of difference in average measurement over a long period of time. In order to keep the gage stable, periodic calibration and maintenance might be necessary. Stability can be illustrated with a control chart, for example, by plotting measurements on the same item over time.

In order to collect accurate process data, we need to periodically test the repeatability and reproducibility of the gages. For the test of bias and stability, we need to periodically calibrate the gages at the certified agencies. Gage R&R doesn't evaluate your product, it evaluates your ability to measure your product accurately and consistently.

Gage R&R calculates the total variation (TV) from three sources:

1. **Appraisers**

2. **Equipment** (e.g., gages)

3. **Parts**

Gage R&R then uses total variation to determine how much variation is attributable to appraisers and equipment (in percentage of R&R). To show how the Gage R&R works, consider the following example from manufacturing.

**Example 6-4: Injection Molding.**

A company produces injection molding products (see Figure 6.13), which is part for cell phones, and the weight of part is a quality characteristic. The target value is 13.86 and tolerance is ± 0.5 (1 total). Each appraiser (A, B, C) measures one sample three times and ten samples are used for the gage R&R study. The data are listed in `Molding.ezs`.

The gage R&R study can be done by following the steps using eZ SPC, and the eZ SPC window is displayed in Figure 6.14, with the results in Figure 6.15.

1. Select the three appraisers labelled A, B, and C.

2. Take 10 parts (conveniently numbered 1–10).

**Figure 6.13** Injection molding for cell phone.

3. For each appraiser and each part, three trials are run.

4. Record the measurement readings.

5. Select `eZ SPC Analysis` ▷ `Gage R&R` and select number of parts, number of appraisers, number of replicates, and type in total tolerance.

6. Type the data in the Gage R&R Design then click `start analysis`.

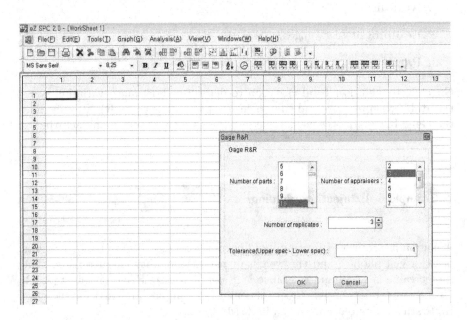

**Figure 6.14** eZ SPC window for Gage R&R analysis.

We will go over all the essential parts of the eZ SPC output, which is shown in Figure 6.16.

- **% Contribution**: This represents the percentage of total variation due to each source. In the output, each value in the Variance column is di-

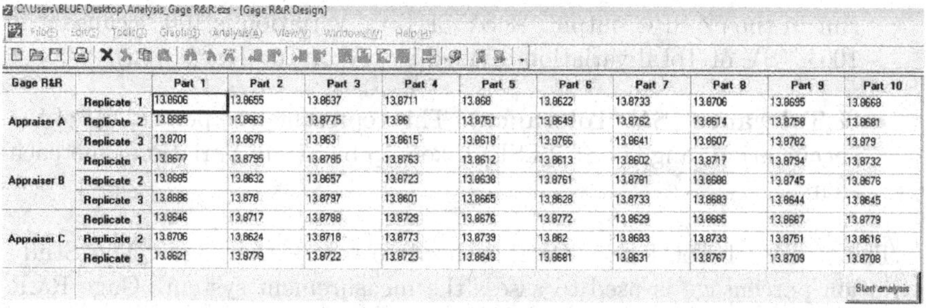

C:\Users\BLUE\Desktop\Analysis_Gage R&R.ezs - [Gage R&R Design]

| Gage R&R | | Part 1 | Part 2 | Part 3 | Part 4 | Part 5 | Part 6 | Part 7 | Part 8 | Part 9 | Part 10 |
|---|---|---|---|---|---|---|---|---|---|---|---|
| Appraiser A | Replicate 1 | 13.8606 | 13.8655 | 13.8637 | 13.8711 | 13.868 | 13.8622 | 13.8733 | 13.8706 | 13.8695 | 13.8668 |
| | Replicate 2 | 13.8685 | 13.8663 | 13.8775 | 13.86 | 13.8751 | 13.8649 | 13.8762 | 13.8614 | 13.8758 | 13.8681 |
| | Replicate 3 | 13.8701 | 13.8678 | 13.863 | 13.8615 | 13.8618 | 13.8766 | 13.8641 | 13.8607 | 13.8782 | 13.879 |
| Appraiser B | Replicate 1 | 13.8677 | 13.8795 | 13.8785 | 13.8763 | 13.8612 | 13.8613 | 13.8602 | 13.8717 | 13.8794 | 13.8732 |
| | Replicate 2 | 13.8685 | 13.8632 | 13.8657 | 13.8723 | 13.8638 | 13.8761 | 13.8781 | 13.8688 | 13.8745 | 13.8676 |
| | Replicate 3 | 13.8686 | 13.878 | 13.8797 | 13.8601 | 13.8665 | 13.8628 | 13.8733 | 13.8683 | 13.8644 | 13.8645 |
| Appraiser C | Replicate 1 | 13.8646 | 13.8717 | 13.8788 | 13.8729 | 13.8676 | 13.8772 | 13.8629 | 13.8665 | 13.8667 | 13.8779 |
| | Replicate 2 | 13.8706 | 13.8624 | 13.8718 | 13.8773 | 13.8739 | 13.862 | 13.8683 | 13.8733 | 13.8751 | 13.8616 |
| | Replicate 3 | 13.8621 | 13.8779 | 13.8722 | 13.8723 | 13.8643 | 13.8681 | 13.8631 | 13.8767 | 13.8709 | 13.8708 |

Start analysis

**Figure 6.15** eZ SPC data for Gage R&R analysis.

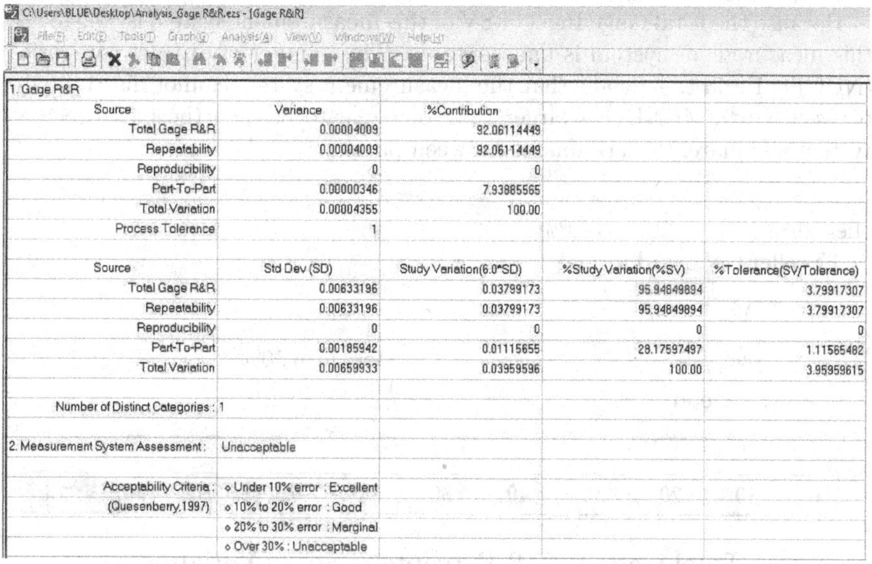

C:\Users\BLUE\Desktop\Analysis_Gage R&R.ezs - [Gage R&R]

| | | | | |
|---|---|---|---|---|
| 1. Gage R&R | | | | |
| Source | Variance | %Contribution | | |
| Total Gage R&R | 0.00004009 | 92.06114449 | | |
| Repeatability | 0.00004009 | 92.06114449 | | |
| Reproducibility | 0 | 0 | | |
| Part-To-Part | 0.00000346 | 7.93885565 | | |
| Total Variation | 0.00004355 | 100.00 | | |
| Process Tolerance | 1 | | | |
| Source | Std Dev (SD) | Study Variation(6.0*SD) | %Study Variation(%SV) | %Tolerance(SV/Tolerance) |
| Total Gage R&R | 0.00633196 | 0.03799173 | 95.94849894 | 3.79917307 |
| Repeatability | 0.00633196 | 0.03799173 | 95.94849894 | 3.79917307 |
| Reproducibility | 0 | 0 | 0 | 0 |
| Part-To-Part | 0.00185942 | 0.01115655 | 28.17597497 | 1.11565482 |
| Total Variation | 0.00659933 | 0.03959596 | 100.00 | 3.95959615 |
| Number of Distinct Categories : | 1 | | | |
| 2. Measurement System Assessment : | Unacceptable | | | |
| Acceptability Criteria : | ◦ Under 10% error : Excellent | | | |
| (Quesenberry,1997) | ◦ 10% to 20% error : Good | | | |
| | ◦ 20% to 30% error : Marginal | | | |
| | ◦ Over 30% : Unacceptable | | | |

**Figure 6.16** eZ SPC output for Gage R&R analysis.

vided by the Total variation then multiplied by 100. Specifically, % Contribution of Total Gage R&R is the sum of three parts: % Contribution of Repeatability, % Contribution of Reproducibility and %Contribution of Part-to-Part.

- **Study Variation** ($6 * SD$): This represents six times the standard deviation of the data from each source.

- **% Study Variation**: Study Variation (SV) represents percentage of Study Variation of each source compared to Total Variation. That is, % SV of Total Gage R&R = 100× (SV of Total Gage R&R divided by SV of Total variation) and all other values are computed in the same way.

But in the eZ SPC output, % SV of Total variation is 100 because it is $100\times$ (SV of Total variation divided by SV of Total variation).

- **% Tolerance (SV/Tolerance)**: This represents the percentage of the specification range (USL - LSL) taken up by Study Variation from each source.

The % SV of Total gage R&R is used to assess the gage. The Study variation percentage is used to assess the measurement system. Gage R&R acceptability criteria are based on the benchmarks provided by Quesenberry (1997), which considers less than 10% to be excellent, and over 30% to be unacceptable. See Figure 6.17.

Because the total Gage R&R % SV of this measurement system is 95.9485%, this measurement system is unacceptable. The number of distinct categories (NDC) is 1 and this means that the measurement system cannot discriminate between parts. If NDC is larger than or equal to 5, then the measurement system adequately discriminates between parts.

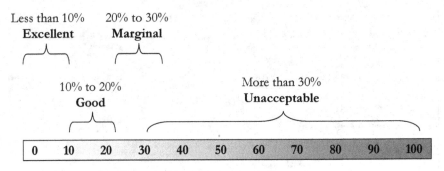

**Total Gage R & R Percentage Study Variation**

**Figure 6.17**    Guidelines for Total Gage R&R % SV according to Quesenberry (1997).

## 6.7   WHAT DID WE LEARN?

- CUSUM and EWMA control charts are constructed to detect gradual changes over time in the process, and differ from Shewhart charts that are made to detect outliers and dramatic changes that occur suddenly to the process.

- The CUSUM and EWMA charts can be used with individual observations.

- The CUSUM procedure uses a set of past data, weighted equally, to plot on the control chart.

- The EWMA procedure uses past data like the CUSUM, but assigns more weight to the more recently sampled outputs.

- The CV control chart is for special situations in which the variance does not remain constant when the process average changes.

- Once the process is stable, we can assess the process capability of a system to find out how many process outputs will conform to specifications.

- Gage R&R studies can help identify the natural variability in the process and distinguish it from measurement error.

## 6.8   TEST YOUR KNOWLEDGE

**1.** Suppose we are constructing a CUSUM chart for a manufacturing process that has an output target value of 5.1 kilograms and standard deviation 1.0 kg. Choose the appropriate reference value $(K)$ for the CUSUM chart.

   a. 0.50

   b. 1.00

   c. 2.50

   d. 5.00

**2.** In Example 3-9, the CUSUM control chart indicated that the process was out of control at a certain point. What was the value of the measurement that triggered the out-of-control alert?

   a. 392 g/2-inches-width

   b. 455 g/2-inches-width

   c. 511 g/2-inches-width

   d. 516 g/2-inches-width

**3.** Again in Example 3-9, at what measurement do we attribute as the start of the detected process shift

   a. 392 g/2-inches-width

   b. 455 g/2-inches-width

   c. 511 g/2-inches-width

   d. 516 g/2-inches-width

**4.** The EWMA chart is based on a weighted average of past measurements, with weights determined by weighting factor $\lambda$. What value of $\lambda$ will produce a control chart equivalent to a CUSUM chart?

    a. 0

    b. 0.5

    c. 1

    d. no value of will produce a chart similar to the CUSUM

**5.** With a weighting factor of $\lambda = 0.2$, the EWMA gives a weight to 4th most recent measurement that is how much larger than the weight given to the 6th most recent measurement?

    a. 10.0 %

    b. 16.7 %

    c. 20.0 %

    d. 56.3 %

**6.** The CV-chart is used to monitor special processes in which

    a. The CUSUM is too hard to use.

    b. A V-mask is applied to the CUSUM.

    c. Variance changes when the mean changes.

    d. The process is in statistical control.

**7.** For a stable process, if the upper specification limit is three standard deviations over the target mean and the lower specification limit is set to be two standard deviations under the target mean, the process capability index is

    a. 0.250

    b. 0.667

    c. 0.833

    d. 1.000

**8.** If $C_p = 1.00$, what proportion of the measured output are considered within specification?

    a. 83.33 %

    b. 90.00 %

    c. 99.73 %

    d. 100.0 %

**9.** Suppose the normal probability plot reveals a noticeably curved pattern in the plotted points. Which of the following statements are true?

    a. The observed data have a distribution with a bell-shaped curve.

    b. The process capability analysis can lead to erroneous results.

    c. The process capability index is larger than one.

    d. The measuring devices should be calibrated.

**10.** In a Gage R&R study, a single output will be measured

    a. Once

    b. Twice

    c. Three times

    d. At least six times

## EXERCISES

**[Exercises 6.1-6.4]** A small brewery makes a Pilsner-style lager and sells them in six packs of 355 ml cans. The net volume specification of a can is 355 +/- 5 (ml). A quality engineer collected data from the process for the process capability analysis. The data are given in `Pilsner.ezs`. The brewer would like to complete a process capability analysis. In order to do this, we must check three assumptions, which we will do in the first three exercises.

**6.1**   (Assumption 1) The data (volume measurements) must follow the normal distribution. We need to check the normality of the data. To do this, select the data and choose `Normality test` under the Analysis tab. What does the Shapiro-Wilk Test result imply? Does this agree with the probability plot you constructed?

**6.2**   (Assumption 2) The data must show that the process is in control. Use the $\bar{x}$-R chart to determine this. Is this assumption satisfied?

**6.3**   (Assumption 3) If the process is deemed to be under control, we still need the process mean to be centered at the midpoint of the specifications. What is the mean according to the $\bar{x}$-R chart? Is this reasonably close to the target value of 355 ml?

**6.4**   (Process Capability Analysis) After the assumptions have been satisfied, proceed with the PCA by selecting the data and then choosing `Process Capability Analysis` under the Tools tab. In this case, we want to enter the USL and LSL values as 360 and 350, respectively, and enter the subgroup size $n = 4$. According to the eZ SPC output, the nonconformity rate from the sample is 0 ppm. What is the potential nonconformity rate? What is the Cpk value? Does the process variation still need to be reduced?

**6.5**   The city water department checks to see if water from the tap has the proper pH balance. For public drinking, the upper limit is 8.5 and the lower limit is 5.5. The city collects twenty 12-ounce water samples on consecutive days, and the data are listed in `CityWater.ezs`. Construct a CUSUM control chart to detect a possible 0.5-$\sigma$ shift of the mean using $h = 5$ and target value $\mu = 7$ for this process. Is the process in control?

**6.6**   Using the city water data from the previous exercise, construct an EWMA control chart using $\lambda = 0.1$ and $L = 2.7$ for this process. Does the result of the EWMA chart agree with the results of the CUSUM chart?

**6.7**   The data in `Transplant.ezs` are partly obtained from Kang et al (2007). Five blood samples were measured to monitor and control the precision of a laboratory's assay procedures. The standard deviation of measurement is approximately proportional to the mean. Construct the CV control chart with the given data. Is this process in control?

**6.8**   The Apple company claims the specification of thickness for their iPhone 4G is 9.3 mm thickness (24% thinner than iPhone 3G). Suppose that the tolerance for the thickness is $\pm$ 0.2 mm and the process data are collected and given in the file `iPhone.ezs`. Do the process capability analysis for this process and explain the results.

**6.9**   A jewelry manufacturing company produces 22-karat piece of jewelry weighing 30 g. The target value is 30 g and tolerance is +/- 0.1 (0.2 total). The chemical lab scale is used to measure the weight accurately. They measured 10 parts with three appraisers. Each appraiser measures a part three times. The data are given in the file `Jewelry.ezs`. Use Gage R & R to determine the quality of the measurement system.

# CHAPTER 7

# PROCESS IMPROVEMENT

### For want of a nail

*For want of a nail, the shoe was lost.*
*For want of a shoe the horse was lost.*
*For want of a horse the rider was lost.*
*For want of a rider the battle was lost.*
*For want of a battle the kingdom was lost.*
*And all for the want of a horseshoe nail.*

This proverb has been repeated since ancient times, and has provided inspiration for princes, paupers and quality improvement teams alike. The poem illustrates the butterfly effect, where small imperfections or defects in a working process can cascade into a callosal failure. The nail might be a minor input factor in a process of military warfare which does not directly affect the outcome of the battle, but can affect other input factors higher up the chain. If we can control the quality of horseshoe nails as well as the shoeing process, we can decrease the probability of losing the horse, thus decreasing the probability of losing the rider, and so on.

*Basic Statistical Tools for Improving Quality.* By Chang W. Kang and Paul H. Kvam  **151**
Copyright © 2011 John Wiley & Sons, Inc.

With any real process, there are both controllable and uncontrollable factors that can affect the quality of the output. As we illustrated above, factors not only affect the output, but they can affect the other factors in the process. We will use statistics to measure the relationship between factors, including whether the potential relationship between any two factors is strong or weak.

### Example 1-3: Renting a Car

From this example in Chapter 1, the process revolves around a customer picking up a car at a rental agency. Several factors in the process will affect the amount of time it takes for a satisfied customer to drive off with a car. They might include

- How long the customer stands in line

- What type of contract is signed by the customer (is it easy to fill out?)

- The agency's car selection process (was the actual car reserved? Does the agent have to call to find out?)

- The experience of the agent (how long does it take to explain supplemental coverage?)

There might be many more factors to consider. These factors not only affect the service time, but they can affect each other. Difficulties turned up in one part of the process can cause other factors to change in a negative way. For example, a complicated contract or an inexperienced agent can slow the process down enough to cause the waiting times of the other customers to increase.

• • •

In this chapter, you will learn how to deal with multiple factors using statistical techniques such as correlation analysis, regression, analysis of variance, and factorial design experiments. The first method, correlation analysis, is the simplest way for us to measure the relationship between two factors.

## 7.1   CORRELATION ANALYSIS

In Section 2 of Chapter 3 (Summarizing Data), we first introduced the idea of *correlation* to describe how two different variables or measurements are related. Now those variables will be representing two different factors, either an input or an output of the process. The input factors can be either controllable or uncontrollable. Using what we learned in Chapter 3, we can describe the relationship between these two factors either graphically (with a scatter plot) or analytically (using a correlation coefficient). If one factor ($X$) increases as the other factor ($Y$) does, then the factors are positively correlated. If $Y$ decreases when $X$ increases, the two factors are negatively correlated. The correlation statistic ($r$) is standardized between $-1$ and $1$, and if $r = 0$, then X and Y are uncorrelated.

## Example 2-1: LCD Panel Quality Data

First, we revisit the scatter plot from Figure 2-14 that is repeated in Figure 7.1. In that example, the temperature of the printer circuit board after assembly (plotted on the horizontal axis) is considered to be a factor that influences distance measurements of the LCD panels. The scatter plot reveals a positive correlation between temperature and distance, and the computed correlation coefficient (provided in the scatter plot) is $r = 0.6485$. This signifies a positive correlation between the two factors. However, just because two factors are correlated, it is not necessarily true that one causes the other. That is, correlation does not imply causality.

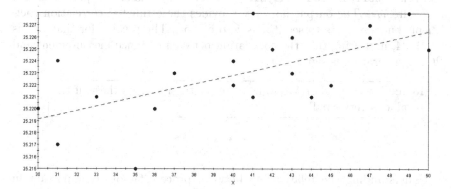

**Figure 7.1**     Scatter plot of distance measurements as a function of PCB temperature after assembly.

In describing how distance and temperature are related, what does the number 0.6485 mean? Does it suggest these two factors are twice as correlated as two other variables that have a correlation coefficient of 0.32425? Actually, this is not the case at all. The correlation coefficient gives you an intuitive way of describing the relationship between the two factors, but the number 0.6485 does not mean they are 64.85% related.

Here is how scientists interpret the correlation coefficient $r$. Statisticians commonly use the square of $r$, denoted by $R^2$, which is called the *coefficient of determination* (see Regression Analysis), to further interpret the relationship between two factors. However, in this case they are assuming that the factors exhibit a linear relationship on a scatter plot. For example, with the PCB temperature data, $R^2 = (0.6485)^2 = 0.4206$ can be interpreted, in a loose way, as saying that about 42% of the variability in one factor can be explained by the other. That is, the distance measurements have an observed variability that can be measured by the sample variance or standard deviation. About 42% of the variance in the distance measurements is due to the fact that the PCB temperature is varying. If the temperatures had remained constant, the distance measurements would have varied much less.

Using eZ SPC, we can compute the correlation coefficient and also test to see if it is significantly different from zero. That is, we can consider $H_0$ : the data are uncorrelated ($r = 0$) and test this hypothesis versus $H_1$ : the data are significantly correlated, which implies there might be some functional relationship between them. To do this correlation test, open the LCD temperature data in `Temps.ezs`. Select both columns of data and under the Analysis tab, choose

`correlation and regression analysis`

and select `Correlation analysis`. The eZ SPC window display is shown in Figure 7.2. The output not only provides the estimated correlation coefficient, but also tests to see if it is zero or not. The p-value for that test is 0.001984, indicating that the correlation between distance measurement and PCB temperature is significant.

---

Reminder for reader: We want to reject the null hypothesis if the p-value is very small.

---

• • •

Understanding the relationship between process factors is important in SPC. If some factors naturally vary together, it can lead to a sub-optimal output. For example, slow service by a car rental agent might cause the line at the rental agency to grow dramatically. These two factors (queue length, agent performance) are two factors that affect service time, and can be strongly interrelated. In this last pairing, the input factor is not controllable, because we cannot set the queue length to see how it affects the agent behind the desk. When input factors are controllable, we can set them at different values to see how the response changes across a broad range of inputs. This kind of procedure can be handled using regression analysis.

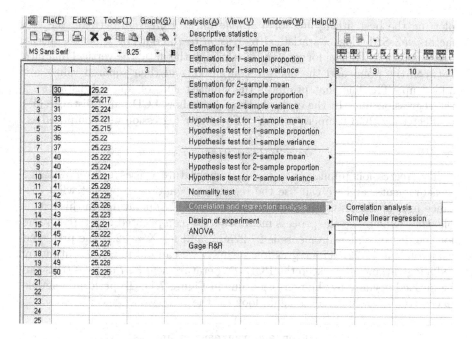

**Figure 7.2**    eZ SPC analysis of PCB data using eZ SPC.

## 7.2   REGRESSION ANALYSIS

Fundamental to process improvement is the idea of finding which inputs matter and how they might be used to optimize the process output. If a controllable input factor is correlated to the output, we have an opportunity to optimize the output by setting the input factor to our liking. To do this, we first need to understand the relationship between the input $(X)$ and the output $(Y)$. *Regression analysis* is used to build an equation relating the output $(Y)$ to one or more input factors $(X)$. With the regression equation that relates X to Y, we can predict the value of output at a new value of the input factor. If there is just one input factor, then the analysis is called a simple regression. Because the relationship is approximated by a line (for example, see the line drawn through the data in Figure 7.1), we specify this as *simple linear regression*. By using this simple framework, we have specified a mathematical equation that describes the output $(Y)$ as a function of the input $(X)$:

$$Y = aX + b.$$

The equation above consists of a slope (a) and an intercept (b), which is the value of $Y$ if we set X to 0. As eZ SPC constructs a scatter plot, it is also plotting the best fitting line for a *simple linear regression* between the two factors. eZ SPC does a statistical regression to the data by what is called a *least-squares technique*; for a given line through the data, the vertical distance between the point and the line are computed, and their squared values are added up. The line that produces the smallest sum of squared distances is the regression line. In the example that shows how LCD panel distance ($Y$) is affected by PCB temperature ($X$). The linear regression was computed as

$$Y = 27.208 + 0.000367X.$$

The regression implies that higher temperature causes greater distance measurements, and we can use the regression equation to predict what distance measurements would likely be given a fixed temperature. In this example, the challenge to the process manager is not complicated; the target value for the distance measurements is 27.22mm, and for X = 32.7, the predicted value of $Y$ is 27.22. The regression suggests we can optimize the output by reducing variance if we keep the PCB temperature near 32.7 degrees (this can be implied visually by the regression line in Figure 7.1). To perform a simple linear regression in eZ SPC, again look for the

### correlation and regression analysis

selection under the Analysis tab. See Figure 7.2. This time choose Simple linear regression. If you select two columns of your data, one column will serve as the input ($X$), and the other column serves as the response predicted by the input ($Y$). Along with the regression equation, eZ SPC generates other statistics and constructs three graphs:

1. **Regression Plot**: This looks the same as the scatter plot in Figure 7.1, which includes the regression line drawn through the data along with the regression equation. You can see more examples of the scatter plot in Section 9 of Chapter 2.

2. **Residual Plot**: A *residual* is defined as the value of the vertical distance between the observed response and the regression line. You can also think of a residual as the difference between the actual output response and what the regression line predicts for any response at that given input. This plot serves as an important diagnostic tool to see if the linear regression is an appropriate tool to use with this particular set of data. We would like to see the residual observations appear scattered across the graph, because if you can spot a distinct pattern such as a curve or a funnel like the plots shown in Figure 7.3, then you have a problem. A distinct pattern serves as evidence that the residuals are correlated. If there is a mathematical relationship between the two factors, we need advanced statistical techniques to describe this relationship; it's too complicated to describe in a simple linear regression.

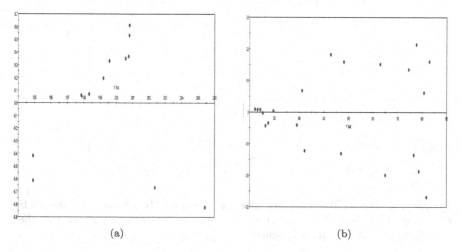

(a)                                                 (b)

**Figure 7.3**   Residual plots that do not display the random scatter we are looking for. Plot (a) shows a curvature, and plot (b) shows a funnel pattern.

Finally, the residual plot also lists a value at the top called R2, which stands for R-square or $R^2$. This corresponds to the squared value of the correlation coefficient in the last section, which was called the coefficient of determination. That means, the $R^2$ can be roughly interpreted as the amount of variability in the output factor that can be explained by predicting it with the linear regression. If the regression produces an $R^2$ value of 0.90, that means 90% of the variance in the output factor is explained by the input factor.

3. **Normal Probability Plot**: Residuals must not only look scattered in the residual plot, but the collected set of residual data must look like it is distributed in a mound-shaped bell-curve (from the normal distribution). Rather than making a histogram of the residual data, the normal probability plot represents a more efficient way of determining this. If the data are normal (with a bell-shaped distribution) then the data should plot approximately on a straight line. Figure 7.4 shows a somewhat straight line, suggesting the data are normally distributed. If there is something wrong with this assumption, you will find curvature and/or a string of outlying points near the end of the line. For more information about the normal probability plot, review the Section *Probability Plots* in Chapter 4 (Analyzing Data).

Along with these three plots, eZ SPC creates a data summary sheet that contains the data, and alongside a column of the predicted output value that corresponds to the given input value. Next to that is a column of residuals,

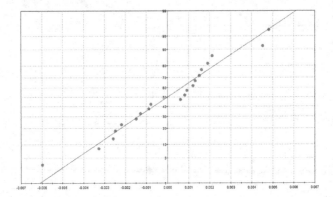

**Figure 7.4**    Normal probability plot for the residuals produced by the PCB regression. There is no detectable curvature, which supports the hypothesis that the underlying data are distributed normal.

which you can verify is just the second column (output) minus the third column (predicted output). Below the data you see a special table of values called an ANOVA (Analysis of Variance) table. We will discuss ANOVA later in this chapter, but one thing we can immediately gain from this table is the p-value, which is listed on the right side of the table. Remember the p-value corresponds to a test of hypothesis (Section 2 of Chapter 4). In this case the hypotheses are simple: the null hypothesis claims the regression is not useful, and the alternative regression claims it is. This can be expressed in terms of the regression equation $Y = aX + b$ by the hypotheses

$$H_0 : a = 0 \qquad \text{versus} \qquad H_1 : a \neq 0.$$

If $H_0$ is true, then $Y = b$, which implies that $X$ does not have any influence on predicting $Y$. So a very small p-value will confirm $H_1$, that the regression is useful (as long as the residual and normal-probability plots are satisfactory).

The following example illustrates how to construct a regression analysis.

### Example 7-1: Grade point averages

We usually don't think about schools as processes that require analysis for quality improvement, but there are several controllable input factors that can be tuned to help the process output (the satisfied student who graduates with their diploma). One particular process output, the students' GPA, is regarded as one of most important factors in their job search. A college admission administrator developed a special test for incoming new freshman students.

She wanted to predict the GPA of new students from the test results, thus allowing the college to gain information on what kind of students will need help in honing their abilities and class performance. She randomly selected 25 freshman students and administered the test. At the end of the academic year, she collected the GPA of those students she tested. The test scores and GPA of each student are given in the file GPA.ezs.

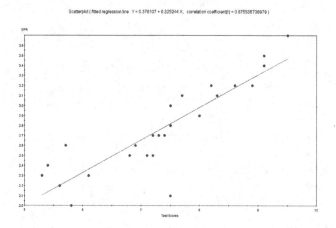

**Figure 7.5**   The scatter plot of test scores (horizontal axis) and predicted GPA (vertical axis), along with regression line and regression equation.

From the scatter plot in Figure 7.5, we see the plotted data, the regression line and the regression equation

$$Y = 0.378107 + 0.325244 * X,$$

where $b = 0.378107$ is the intercept and $a = 0.325244$ is the slope. The slope tells us that when the test score increases one point, we expect the student's GPA to increase 0.325244. The slope is noticeably positive, as students with higher test scores tend to achieve higher GPA on the average.

Suppose the test is administered to a new group of incoming freshman, and one student's test score is 8.8. According the regression, his predicted GPA at the end of year could be

$$0.378107 + 0.325244 * 8.8 = 3.240254.$$

In other words, for students who achieve a test score of 8.8, their expected GPA at the end of year is 3.24.

The information provided in Figures 7.5 and 7.6 help to verify that the regression was appropriate for the data. Although most of the data seem scattered in a random manner, there is one point on the bottom of the residual plot in Figure 7.6 that sticks out - it doesn't fit in well with the rest of the data. If you check the data sheet, your natural sleuthing skills will lead you to determine this is student 10, who scored 7.5 on the test. Although it was predicted this student would get a GPA of 2.8, the actual GPA at the end of the freshman year was only 2.1. This is a good example of an *outlier*, and in this kind of example, the college might find it worthwhile to check in with this student to see if remedial help is needed or if extenuating circumstances ensued.

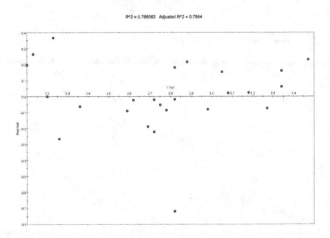

**Figure 7.6**   Residual plot for GPA regression data, along with $R^2$ value.

Below is the eZ SPC computer output that (under the Data Tab) that goes along with the residual plot presented in Figure 7.3. In the first two columns, you will see the original data from the file `GPA.ezs`. In the next column, labelled `Y hat`, are the values of the regression line to correspond with each value in the X column. For example, the first entry of $X = 8.3$ refers to a student with a test score of 8.3 who achieved a grade point average of 3.1. For this student, `Y hat` is 3.0776, which predicts a grade point average of 3.0776 for any student with a test score of 8.3. In this case, the difference between what the student achieved (3.1) and what the regression predicts (3.0776) is quite small, and is listed in the last column, labelled `Residual`: $3.1 - 3.0776 = 0.0224$.

| X | Y | Y hat | Residual |
|---|---|-------|----------|
| 8.3 | 3.1 | 3.0776 | 0.0224 |
| 5.4 | 2.4 | 2.1344 | 0.2656 |
| 7.5 | 3 | 2.8174 | 0.1826 |
| 7.3 | 2.7 | 2.7524 | -0.0524 |
| 5.6 | 2.2 | 2.1995 | 0.0005 |
| 9.1 | 3.4 | 3.3378 | 0.0622 |
| 7.2 | 2.5 | 2.7199 | -0.2199 |
| 5.3 | 2.3 | 2.1019 | 0.1981 |

```
6.9  2.6   2.6223   -0.0223
7.5  2.1   2.8174   -0.7174
9.5  3.7   3.4679    0.2321
8.6  3.2   3.1752    0.0248
6.8  2.5   2.5898   -0.0898
5.7  2.6   2.232     0.368
6.1  2.3   2.3621   -0.0621
5.8  2     2.2645   -0.2645
8.9  3.2   3.2728   -0.0728
7.2  2.7   2.7199   -0.0199
7.5  2.8   2.8174   -0.0174
7.7  3.1   2.8825    0.2175
8.2  3.2   3.0451    0.1549
7.1  2.5   2.6873   -0.1873
8    2.9   2.9801   -0.0801
9.1  3.5   3.3378    0.1622
7.4  2.7   2.7849   -0.0849
```

```
ANOVA Table
Source          Sum squrs(SS) DF  Mean squr(MS) F0      p-value
Regression      3.7365         1  3.7365        75.528  0.0000000070671
Residual Err    1.1379        23  0.0495
Sum             4.8744        24
```

```
(significance level 5\%)  Reject H0
(significance level 1\%)  Reject H0
```

The additional information provided by eZ SPC in this problem (starting with the title ANOVA Table can help with more advanced regression analysis, but is not necessary for us to understand the fundamental outcome of the GPA analysis. You can see the familiar p-value in the far right portion of the table (p-value = 0.000000007067115) which corresponds for the test of $H_0$ : the regression is not useful versus $H_1$ : the regression is helpful. If the p-value was larger than 0.05, we would have to conclude that test scores are not sufficiently helpful in predicting GPA. In this case, though, the p-value is incredibly small (less than one in one hundred million) and the data strongly support the alternative hypothesis that states $X$ and $Y$ are related.

In Figure 7.7 we see the probability plot for the GPA data. Admittedly, this is no straight line by any means, but neither is it decisively curved. Undoubtedly, the observation on the far left (which also pertains to student 10) has had an affect on the plot. Overall, the outlier is notable, but there is no strong evidence that the residual distribution is not normal (bell-shaped).

The normal probability plot is a subjective way of deciding whether or not the assumptions about the residuals are valid. If we construct a histogram for the set of residual data, its distribution should also appear bell-shaped, like a normal distribution. Look at the histogram of the GPA residual data in

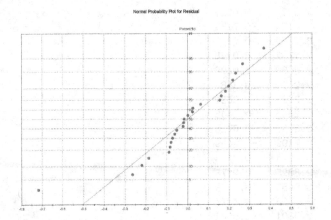

**Figure 7.7**   The normal probability plot for the GPA regression data.

Figure 7.8. There are an insufficient number of observations for a histogram to really show us what the real underlying distribution of the data will look like. We can see a vague mound shape that characterizes the normal distribution, and we also see the effect of the outlier in the right-most bin of the histogram.

● ● ●

In summary, regression is useful in learning about how controllable factors can be used to optimize or control process output. For uncontrollable factors that cannot be preset in an experiment, the relationship between the input and output is more easily interpreted using a correlation coefficient.

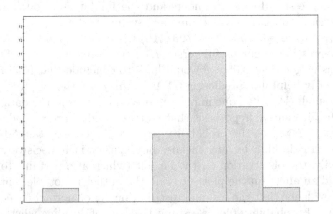

**Figure 7.8**   Histogram for the residual observations that were generated by the eZ SPC regression procedure.

## Regression to the Mean

In 1886, Sir Francis Galton observed in people that extreme characteristics (such as height) in parents are not always passed on to their offspring. Using statistical regression, he noticed "the average regression of the offspring is a constant fraction of their respective mid-parental deviations". When Galton noticed how the childs height differed from the height of his parents, the childs height tended to be taller if the parents were unusually short and shorter if the parents were unusually tall.

This phenomenon, called *Regression to the Mean*, occurs everywhere in nature and frequently leads people to attribute a measured change to some prescribed cause. For example, U.S. Navy flight trainers would chastise pilots who performed poorly on a flight test and thought this was an effective coaching method. Data showed that on the subsequent flight, pilot performance improved. But the improvement need not be attributed to being yelled at. After any unusually poor performance, flight test scores are due to regress to the mean, which will be a higher score. Trainers who believed their methods of harsh discipline caused pilots to fly better were deceived by this regression fallacy.

Regression to the mean helps explain the performance of the stock market, the variability of sports outcomes, and even medical diagnostics. When a sufferer of chronic pain has an unusually distressing day, the pain might cause him to seek medical attention. If the pain subsides the next day, the patient is likely to believe the medical treatment was the cause of improvement. In truth, it might be more attributable to the day-to-day variation in suffering that the patient experiences.

Consider how regression to the mean can affect the way we perceive process improvement using a control chart. If a process is in control, our control chart should have a large average run length $(ARL)$. On those rare occasions when a measurement falls outside the control limits, we know to look for an assignable cause that will explain why the process was deemed to be out-of-control. If the process is actually in control, however, it might be hard to find any kind of assignable cause, and if the wrong cause is attributed to the deviation in this observation, it will be reinforced once we observe future output measurements that are back within the control limits.

## 7.3  EXPERIMENTAL DESIGN

With any set of factors that influence the output of a stable process, the process manager can fiddle with the system by changing the level of one of the factors now and then to see what level of that factor works best in creating the highest quality output. This philosophy of adjusting one factor at a time is a natural way to experiment and find optimal solutions to complex processes as well as our daily tasks. For example, we might suspect that most recipes in a well-liked cookbook have benefited by this kind experimentation. However, this one-factor-at-a-time way of experimenting is not always the best way, notably in industrial processes. We consider a couple of examples, starting with the story of Goldilocks and the Three Bears.

## Example 7-2: Goldilocks and the three bears

In a well known fairy tale, Goldilocks is a little girl who conducts an experiment. She has spotted a cottage where three bears live. One day, the bear family temporarily leaves their quiet abode in order to let their lunch-time porridge cool, Goldilocks enters the house. Her goal is to determine factor levels that optimize her comfort in regard to sitting, eating and resting. She starts her experiment by finding the coziest chair in which to sit. She experiments with all three, and after finding the father bear's chair too hard, the mother bear's chair too soft, Goldilocks decides the baby bear's chair is just right. Unfortunately, during the experiment, the baby bear's chair is broken. Goldilocks then conducts a taste test with the porridge, finding the father bear's porridge too hot, the mother bear's porridge too cold, and the baby bear's porridge just right. Goldilocks eats up all of the baby bear's porridge, although it is not clear this is a necessary part of the experimental protocol.

Next, to optimize her resting comfort, she next tries out all three bear beds, finding the father's bed too hard, the mother's bed too soft, and the baby's bed just right. After experimenting with different levels of the three factors (chair, porridge, bed), she falls fast asleep. Technically, Goldilocks' designed experiment has ended in success, although the fairy tale continues when the bears return home. The baby bear, of course, is rightly upset after noticing the broken chair and missing porridge. In some versions of the story, the bears eat Goldilocks, which makes us rethink what we define as a success in her experimental outcome. Perhaps the three comfort measurements are not sufficient in measuring success; being comfortable does not hold as much meaning if you end up being eaten.

• • •

In real-life processes, we cannot always find the best process output by fiddling with one factor at a time. For example, if Goldilocks had used a less comfortable chair that would not fracture, her immediate comfort might decrease, but perhaps the bear family would have inflicted a less harsh penalty knowing all their bear furniture was intact. The next example considers two related input factors that determine the process output (driving time), but this example does not contain bears.

### Example 7-3: Minimizing driving time

To minimize the time used to drive to the office, a commuter considers two routes to work (interstate, side roads) and two different times to start the commute (7:30 AM, 8:30 AM). All four route combinations are repeated five times over the course of four work weeks. The 7:30 trip using the interstate averaged 34 minutes, while the later 8:30 AM trip averaged 32 minutes. The 7:30 side road commute takes an average of 44 minutes, and the 8:30 trip on the side roads takes 36 minutes, on average. Overall, the Interstate provided the faster route (average 5 minutes faster than the side roads), and leaving earlier brought the commuter to work 5 minutes quicker on average. But the best combination is not to take the interstate at 7:30, because it was shown that the later commute on the interstate is two minutes quicker.

|         | Interstate average | Side Road average |
|---------|--------------------|-------------------|
| 7:30 AM | 34 minutes         | 44 minutes        |
| 8:30 AM | 32 minutes         | 36 minutes        |

• • •

The commuter's experiment has a big advantage over Goldilocks' experiment. The commuter was allowed to change factor levels simultaneously, allowing us to see the best combination is not determined by the average effects of the two factors of road type and starting time, which imply earlier rides and interstate driving is optimal. The commuter generated a well-designed experiment. Goldilocks did not have this luxury, so while the commuter has happily minimized the time it takes to arrive at the office, Goldilocks was viciously consumed by three anthropomorphic bears.

## 7.4 OVERVIEW OF EXPERIMENTAL DESIGN

Although we have a good instinct at experimenting by changing the settings of a process, experimental design is meant to find the best combinations of settings that will help us achieve our goals in an efficient and affordable manner. We looked at examples with two or three factors, each with just a few factor settings. This limited scope might allow us to repeat the experiment at every conceivable factor level without prohibitive cost. But many processes will have numerous factors to consider, and numerous factor levels for each factor. If the process output depends on 10 input factors, and each input factor can be set to four different levels, we would have to repeat the experiment over a million times ($4 \times 4 \times 4 \times 4 \times 4 \times 4 \times 4 \times 4 \times 4 \times 4 = 4^{10}$ to be exact) to achieve an output at every combination of factor level. For many actual processes, it might be more realistic to think the experiment could be repeated between 10 and 30 times. There is just too much expense and labor invested in an experimental run that prohibits the process manager from repeating the experiment indefinitely.

The design of an effective experiment means finding out what input factors are important, what factor levels to choose from these factors, and what combination of these factor levels will give us the best output from the process. The objectives of the experiment depend on the process; we might want to optimize the output by raising the target mean, decreasing the variability, or both. Three general objectives for an experiment are the following

1. Selection of a set of input factors that seem to be the most influential on the process output in terms of process quality.

2. Setting the values of an input factor in such a way to make the output close to the target value.

3. Setting the values of the input factor in order to make the process output less variable.

A well designed experiment is the most important step to achieve the most information from the experiment with the least amount of investment. It will be vital to learn which factors are important, but sometimes it is just as important to find out which ones are not influential on the process.

The design of an experiment starts long before data are sampled. From identifying the objectives to realizing the cost savings that can be provided by a designed experiment, the guidelines can be described in the following steps:

**1. Identify and understand the problem:** Is the process out of control, or do we want to increase the output for an in-control process? What factors influence the output? Which ones are controllable?

**2. Select the response measurement:** What are you measuring from the process output? In some industries, this might be obvious (counting defects, etc.) but be careful what you measure in the output to determine its quality. Don't make the mistake that Goldilocks made!

**3. Select the controllable factors:** Set the levels of each factor for the experiment. Usually, experimenters try to minimize the number of factor levels used in an experiment (perhaps to two levels) in order to reduce the dimension and cost of the designed experiment. Make sure the factor levels are set to values that will be used in practice.

**4. Select the particular way the experiment will be designed:** We will cover some basic designs in this chapter. A design for a single input factor is introduced in Section 4, and two-factor designs are outlined in Sections 5 and 6.

**5. Select the experimental units to be sampled:** Which process outputs should be monitored? Should the experiment be run consecutively? Or should it be spread out? Will the sampling hold up the production process? We will discuss how randomness is used in selecting experimental units.

**6. Perform the designed experiments:** Make sure everything is done according to plan. If there is improper sampling or missing data, the validity of the experiment may be dubious.

**7. Collect and analyze the data:** Most analysis is computationally cumbersome, so we rely on statistical software such as eZ SPC to do the dirty work for us. Along with the computing-intensive analyses, graphical methods are also a vital part of the data analysis.

**8. Draw a tentative conclusion:** This might include the interpretation of confidence intervals or hypothesis tests with p-values.

**9. Perform the validation experiments:** In many cases, the experiments were run in the laboratory, which is a well-controlled environment, so the results of the experiments may not be reproduced if the same experiments are run inside a manufacturing plant or out in the field. It is necessary to run validation experiments in order to make sure that the experimental results can be applied to real-life situations where the environment is less controlled.

**10. Draw the final conclusion and recommend improvements:** After reviewing the results of the tests, the experimenter can confidently report the results of experiments and recommend improvements, with a suggested change of input factor levels if necessary.

## Example 7-4: Semiconductor Manufacturing

In a semiconductor manufacturing plant, one etching machine started to produce more frequent defects on wafer surfaces to such an extent that the process was deemed out of control. In order to find out why this etcher was fabricating defective wafers while the other machines were stable, the lab group held brainstorming sessions to share expert knowledge of the fabrication process, the input of raw materials, and the operating environment. The group decided to focus on three potential problem areas: the power transistors, the pressure level of the etch chamber and the combination of reactive gases (fluorocarbons, oxygen, chlorine, boron trichloride) used in the etching.

To find out which of these three controllable factors were influencing the quality of the output, an experiment would be run using different combinations of factor levels for the three inputs. Experiments are run at both high and low power levels, and with high and low pressure, so this makes four different combinations to study. For the third factor, three levels of oxygen (in the reactive gases) are used in exposing the material to a bombardment of ions, so that the experiment has to be repeated $4 \times 3 = 12$ times. Based on what they had for available resources, the plant operators repeated the experiment at each level three times, so 36 experiments were run, in total. Once all the data are in, the laboratory wants to set the etching process at the three factor levels that produced the fewest defects.

$$\bullet \quad \bullet \quad \bullet$$

## 7.5   PRINCIPLES OF EXPERIMENTATION

It's important to remember that experimental design occurs before any data are collected. Design does not imply data analysis, but rather it represents a plan for making the analysis as efficient, cost effective and easy as possible. The experimental design will provide instructions for setting the input factors and sampling process output. There are three basic principles of a good design: randomization, replication and blocking. The first two boil down to common sense, but the third principle of blocking actually contains a fair amount of uncommon sense.

1. **Principle of Randomness**. In order to get objective data from the experiment, randomness takes an important role in whole processes of experimentation. Randomness should be guaranteed in selection of experimental units, assignment of treatment to experimental units, and order of experiments. The experimental units are randomly selected from the pool of resources. The treatment for experiment should be assigned to the experimental units at random. When the experimenter decides the order of experiments, he or she should use the method of random number generation.

2. **Principle of Replication**. By repeating the experiment, the process manager can obtain an improved estimate of output variability. With more replication, the experimental conclusions become stronger. For each combination of factors that characterize a single experimental run, at least three to five replications are recommended.

3. **Principle of Blocking**. In experimentation, some factors may be crucial in achieving the best output, but cannot be controlled during the

day to day process. For example, a process output might vary greatly because of the effect of the person involved in controlling or managing inputs during the process. If the process output has higher quality when Worker A is on duty, compared to when Worker B is on duty, we should be wary of conclusions about how the other factors influence the output. Instead of merely firing Worker B, we can increase the efficiency of the experiment by insisting that each worker is on line for experiments involving every combination of factor level, ensuring that the entire experiment is repeated twice.

Not only does the data help us contrast the effect of the two workers, but now the effect of the workers can be separated from the effects of the other factors. We are blocking by worker, and now the conditions for the experimental units will be similar and the conclusions more fair. In general, the experimenter needs to make the conditions of experimental units as similar as possible. By assigning similar experimental units into same block (e.g., worker), the experimenter may remove the ambiguity in interpreting the relationship of factors. If there is any relationship between the factors, we know that it resulted from the characteristics of those factors, and not the blocking factor.

In the next three sections, we will introduce you to three basic experimental designs that can help you conduct efficient experiments. They involve experiments for one factor (Section 6) and two factors (Sections 7 and 8), which can generalize into experimental designs for three or more factors. Each design leads to a statistical summary called *Analysis of Variance*, which is sometimes shortened to ANOVA.

Analysis of Variance describes experimental investigation in which the observed variance of the process outcome can be partitioned into different distinct parts due to different explanatory variables. For example, if there are four different controllable factors that affect the outcome of an experiment, they also contribute to the variance of the experimental outcome, and by partitioning the variance, we mean to assign a certain amount of cause for output variation to each of the factors involved. Less important factors will not affect output as much as others, and the component of variability assigned to that factor will be less. The first ANOVA we will go over has only one factor, so this partitioning is not as meaningful. When we have two or more factors (Sections 7 and 8), then splitting the variance of the output into separate components will be more consequential. The concept of partitioning variances is an abstract one, and we will illustrate the idea in Example 7-5 in the next section.

## 7.6   ONE-WAY ANALYSIS OF VARIANCE

A one-way analysis of variance implies that we are focused on just one input factor that affects the process output. For this single factor, several factor levels can be chosen when undergoing the experiment, but the experiment must be repeated at each of the chosen levels. Statisticians call the different levels of the factor the *treatment levels*, probably due to the agricultural origins of these statistical methods and the frequent use of fertilizer treatments in crop experiments. The purpose of one way analysis of variance is to determine whether all treatments have same effect on output or not.

## Example 7-5: UV Radiation

To study the effects of UV (ultraviolet) radiation, experimental mice are subjected to varying levels of UV intensity. Four different levels of UV index (UVI) levels are chosen (7, 9, 11, 13) because they are important for study and potential extrapolation to human susceptibility. Exposure time is set to 30 minutes. Three mice will be randomly assigned for each treatment, so 12 mice are randomly selected for the experiment. To ensure the experiment is properly randomized, a random number generator is used to assign the 12 mice to the four different treatment levels. After 12 runs of the experiment, the experimenter will count the number of healthy cells in a pre-specified area of each mouse's skin. The data are located in the file UV.ezp.

In Example 7-5, we have a classic one-way ANOVA with four treatment levels and three observations at each treatment level. This design is called a *Completely Randomized Design* because of how the mice were assigned to treatment groups. When the experiment is finished, we compare the treatment means to find out if there are any differences among them. Even though we call this an analysis of variance, it seems more akin to an analysis of means. The "variance" part is there because we are going to assume the samples at each UV level have a similar amount of variability. That means that even though we have four samples and four sample means, we are not going to estimate four sample variances (see $s^2$ in Section 2 of Chapter 3). Instead, we combine information from all of the treatment groups and make an estimate of an overall variance, and this is the key to ANOVA for the one-way analysis. To keep our chapter lesson simple, we refrain from describing the details about how this combined variance is estimated from the data.

To find out if the treatment means are different, we can use the hypothesis testing framework from Section 2 of Chapter 4. In the example, we initially want to find out whether the average number of healthy cells is equal or not at different the four different levels of UV. The null hypothesis will be the equality of the treatment means. The alternative hypothesis is the broad conjecture that at least one mean level is different from others.

$H_0$: The average number of healthy cells is the same for all four UV levels.

$H_1$: For at least one of the UV levels, the true average number of healthy cells is different from others (but no specific alternative truth is conjectured yet)

To start the analysis of variance in eZ SPC, first enter the data in a single column, preferably the second column on the spreadsheet. In the first column, alongside the observed count, enter the UV level (7, 9, 11, 13). After selecting both columns of data on the spreadsheet, now go to the Analysis tab and select ANOVA (at the bottom of the Analysis tab), and then choose

`One Way ANOVA.`

Make sure the columns are properly matched up (input factor versus response) in the `select variables` window.

We will not scrutinize the details of the test statistics and the computational forms for the ANOVA. In short, the four sample means are compared based on the assumption that they have the same underlying variance. If you check the chart in Figure 7.9, created by eZ SPC, the four treatment means are plotted along with 95% confidence intervals for those mean values. From the figure, we can see how the first two treatment levels (UV = 7, 9) appear to produce significantly higher mean counts than the third (UV = 11), which itself seems significantly higher than the fourth (UV = 13).

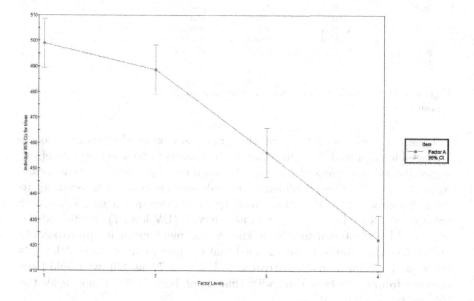

**Figure 7.9**    Chart of treatment means at four different IU levels. Factor level 1 refers to UV level 7, factor level 2 refers to UV level 9, and so on.

In Figure 7.10, the ANOVA table from eZ SPC is reproduced. The ANOVA table represents a complete statistical summary of the experiment outcome. Again, we will spare you the computational details in the analysis, but we will emphasize the important things you can learn from this table. In part 1 (ANOVA Table), the first thing to look at is the last column on the right which lists a p-value (0.00000379). This p-value corresponds to the basic hypothesis test we described above, so the evidence emphatically points to the alternative hypothesis, which states that the mean counts are not the same. This is an analytic justification for what we visually determined in Figure 7.9. The p-value is based on an *F-test* that compares the variance of the data when each treatment group has a unique mean (our model) versus the variance of the data when no such treatment effects are assumed.

| 1. ANOVA Table | | | | | |
|---|---|---|---|---|---|
| Source | Sum of squares(SS) | DF | Mean square(MS) | F0 | p-value |
| Factor | 10804.916667 | 3 | 3601.638889 | 72.760382 | 0.000003789962946 |
| Error | 396 | 8 | 49.5 | | |
| Sum | 11200.916667 | 11 | | | |
| Test result : | (significance level 5%) Reject H0 | | | | |
| | (significance level 1%) Reject H0 | | | | |

| 2. A 95% CI on the treatment mean | | | | |
|---|---|---|---|---|
| Level | Lower confidence limit | Mean | Upper confidence limit | Replicates |
| 7 | 489.393325 | 499 | 508.606675 | 3 |
| 9 | 479.059991 | 488.666667 | 498.273342 | 3 |
| 11 | 446.726658 | 456.333333 | 465.940009 | 3 |
| 13 | 412.726658 | 422.333333 | 431.940009 | 3 |

| 3. A 95% CI on the difference in two treatmens means | | | |
|---|---|---|---|
| Level | Lower confidence limit | Mean | Upper confidence limit |
| (7 - 9) | -3.252557 | 10.3333 | 23.919224 |
| (7 - 11) | 29.080776 | 42.6667 | 56.252557 |
| (7 - 13) | 63.080776 | 76.6667 | 90.252557 |
| (9 - 11) | 18.747443 | 32.3333 | 45.919224 |
| (9 - 13) | 52.747443 | 66.3333 | 79.919224 |
| (11 - 13) | 20.414109 | 34 | 47.585891 |

**Figure 7.10** ANOVA table for the one-way analysis of variance based on UV exposure data.

In part 2, labeled A 95% CI on the treatment mean, the treatment means are listed for all four UV levels, along with the upper and lower bounds of the respective 95% confidence intervals. We could compare the treatment means by seeing whether these confidence intervals overlap, but it will be easier to learn this in part 3 (A 95% CI on the difference in two treatment means). For example, the sample mean for treatment level 1 (UV level 7) is 499, and the 95% confidence interval for the unknown treatment mean is approximately (489.4, 508.6). If someone had claimed that the mean number of healthy cells found on a specimen that was treated to UV level 7 was 475, we would have evidence from the data to contradict that claim, because 475 is not inside the 95% confidence interval.

In part 3, we are presented with six pair-wise comparisons between all of the treatment means, including the mean difference of the sample counts along with the 95% confidence interval for the difference. For example, the first row

(7 − 9) lists the mean counts from UV level 7 minus the mean counts from UV level 9. The positive mean value (10.333) tells us that the mean for the UV-7 group is 10.33 units larger than the mean for the UV-9 group. The 95% confidence interval for the mean difference is roughly (−3.25, 23.92). With this interval, it is critical to notice that the number zero is in the interval (though not by much). Since the confidence interval reflects what mean differences can be considered plausible, so if it is claimed that there is no mean difference between the UV-7 and UV-9 group (so the mean difference is zero), then the ANOVA reports that there is not sufficient evidence in the sample to dispute this claim.

Although the UV-7 group and UV-9 group are not considered significantly different, notice that from the other five pair-wise comparisons, all of the remaining treatment pairs are considered to be different. This outcome can be intuitively understood by looking at the confidence intervals in Figure 7.10. Notice, for example, that only the two left-most intervals overlap each other, suggesting that they might share a mean difference.

• • •

## 7.7   TWO WAY ANALYSIS OF VARIANCE

Two-way analysis of variance extends what we are doing with one factor in the one-way ANOVA to an ANOVA with two factors. That is, we are focused on two different input factors that affect the process output. For each of the two factors, several factor levels can be chosen for the experiment, and the experiment must be repeated at each combination of chosen levels. In a single run of the experiment, a level is set for each factor. If there are three levels of the first factor, and five levels of the second factor, there are 3x5 = 15 combinations of the two factor levels.

Let's return to Example 7-3, where two factors (Road, Start Time) are used to predict the experimenter's morning commute time. Both factors have two levels, so there are four driving combinations in all. On average, it was found that the interstate highway was a quicker route to get to the office, and the earlier leaving time (7:30 AM) also leads to a faster commute. These are the *main effects* of the two factors. However, the best overall combination was to take the interstate, but at a later time (8:30 AM). That means we cannot predict the best outcome by relying on the average (or main) effects of the individual factors. This means the experimental affect of one of the factors actually depends on what is going on with the other factor. This is called a factor *interaction*, or an interactive effect. It's going to make things harder to predict knowing we cannot merely tweak one factor at a time.

With a single factor in the experiment, we first analyzed the data to determine if the factor levels even mattered in determining the process output.

Once it was determined the factor was significant, we looked further into the data to find out which factor level produced the best output. With two factors, we again need to determine if either one or both of the factors significantly affects the process output. This is a fairly straight-forward extension to the one-way ANOVA. However, along with testing the factors individually, we also need to determine if the two factors interact significantly. Now there are three basic hypotheses to be tested:

**Null hypothesis 1**: For Factor 1, the means of different levels are equal

**Null hypothesis 2**: For Factor 2, the means of different levels are equal

**Null hypothesis 3**: There is no interaction between two factors

Next we will see how to analyze a two-factor ANOVA using eZ SPC.

## Example 7-6: Tests of throwing distance

A high school physical education teacher wants to examine differences in a student's throwing distance for a baseball (in meters) in relation to sex (M = male, F = female) and training period (T0 = no training, T1 = one month training, T2 = two months of training, T3 = three months of training). The teacher randomly selected three students from each of the eight different combinations of factor levels: F-T0, F-T1, F-T2, F-T3, M-T0, M-T1, M-T2, and M-T3. Each student threw a baseball and the following distance measurements were recorded:

|     | T0         | T1         | T2         | T3         |
|-----|------------|------------|------------|------------|
| F   | 12, 14, 15 | 21, 19, 23 | 29, 32, 31 | 37, 36, 36 |
| M   | 28, 29, 18 | 33, 37, 39 | 40, 38, 41 | 50, 43, 49 |

To run the experiment, you can use one of the two following ways. First, you can have eZ SPC set up the ANOVA data spreadsheet by looking at the bottom of the Analysis Tab, select

```
Design of Experiment
```

from the menu, and then

```
Two Factor Analysis
```

from its sub-menu. eZ SPC creates a spreadsheet with the exact number of factor levels that you specify for each factor. Once you finish filling in the table, click on the red `Start Anova` button.

Another way to construct the two-way analysis of variance is to enter the data on a regular spreadsheet, using three columns. The first two columns represent identifiers for the factor levels, and the third column contains the experimental response (throwing distance) corresponding to an experiment run with those two levels. Using this method, you will have a spreadsheet with three columns and 24 rows. When you select ANOVA under the Analysis tab, next select

`Two − Factor  ANOVA`

and make sure the columns you select match the factors in your experiment.

The computer output consists of two new windows (Chart, Data). The chart is shown in Figure 7.11, and plots the main effects of the two factors. The top line represents the mean throwing distance for the two sexes (1 = Female, 2 = Male). The lower line traces the mean throwing distance (averaged over men and women) for each level of training, from 0 hours to 3 hours (1 = 0 hours, 2 = 1 hour, 3 = 2 hours, 4 = 3 hours).

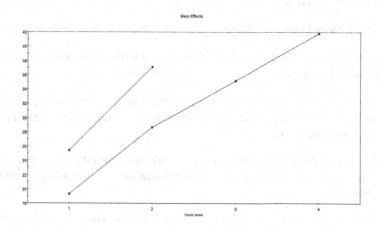

**Figure 7.11**    Graph of main effects for two factors in ball-throwing experiment: sex (male, female) and hours of training (0,1,2,3).

To understand the results, look at the ANOVA table in Figure 7.12. The sheet lists three sections of information just like we saw for the one-way ANOVA:

```
1.      ANOVA Table
2-1.    (Factor A) - A 95\% CI on the treatment mean
2-2.    (Factor B) - A 95\% CI on the treatment mean
```

3.    (AB Combination) - A 95\% CI on the treatment mean

In part 1, the ANOVA table lists four rows that represent the different sources of variability in the response. Source A represents the effect of the first factor entered in the computer analysis, which in this case was sex. Source B symbolizes the effect of the second factor, training hours. The source labeled A*B represents the potential interaction between these two factors. The final error source is all the variability in the response that has not been explained by the other three factors, so this corresponds to noise from the uncontrollable factors in the process.

The ANOVA table contains a lot of miscellaneous information about these four sources of variance, but we will hone in on the critical information that we need to understand the outcome of the analysis. Look at the right most column of data in the ANOVA table that lists p-values. The table now lists three p-values instead of just one. Recall, for the two-factor analysis, we have three tests to consider:

**Null hypothesis 1**: The mean distances for both sexes are equal

**Null hypothesis 2**: The mean distances for all four different training periods are equal

**Null hypothesis 3**: There is no interaction between the two factors

The p-values correspond directly to these three hypotheses, and they lead to the following conclusions for the ball-throwing data:

- We reject the first null hypothesis because the p-value is incredibly small, so the data show the difference between male and female throwing outcomes is significant.

- We reject the second null hypothesis because the p-value is also incredibly small, so the data show the amount of training affects throwing outcomes.

- Finally, we do not reject the hypothesis that there is no interaction between the sex factor and the training factor (please excuse the double negative!). The p-value of 0.35 is not small enough to consider the interaction significant. Recall from Chapter 4, the p-value needs to be less than 0.05 to provide strong evidence to reject the null hypothesis.

In part 2 of the ANOVA output, 95% confidence intervals are computed for the average effect of each of the two sex combinations and four training levels. If a significant interaction existed between the two factors, then we could not count on the factor level means from part 2 to completely describe their combined effects on throwing distance. The main effect of sex is simply the difference between the means of female and male students, for example,

| 1. ANOVA Table | | | | | |
|---|---|---|---|---|---|
| Source | Sum of squares(SS) | DF | Mean square(MS) | F0 | p-value |
| A | 816.666667 | 1 | 816.666667 | 90.740713 | 0.000000053766979 |
| B | 1656.166621 | 3 | 552.05554 | 61.339486 | 0.000000005394191 |
| A*B | 31.666668 | 3 | 10.555556 | 1.172839 | 0.351048191027047 |
| Error | 144.000044 | 16 | 9.000003 | | |
| Sum | 2648.5 | 23 | | | |
| Test result : | (significance level 5%) Reject H0 | | | | |
| | (significance level 1%) Reject H0 | | | | |
| | (significance level 5%) Reject H0 | | | | |
| | (significance level 1%) Reject H0 | | | | |
| | (significance level 5%) Fail to reject H0 | | | | |
| | (significance level 1%) Fail to reject H0 | | | | |
| 2-1. (Factor A) - A 95% CI on the treatment mean | | | | | |
| Level | Lower confidence limit | Mean | Upper confidence limit | Replicates | |
| 1 | 23.571167 | 25.416667 | 27.262167 | 12 | |
| 2 | 35.237833 | 37.083333 | 38.928833 | 12 | |
| 2-2. (Factor B) - A 95% CI on the treatment mean | | | | | |
| 1 | 16.723402 | 19.333334 | 21.943265 | 6 | |
| 2 | 26.056735 | 28.666666 | 31.276598 | 6 | |
| 3 | 32.556735 | 35.166667 | 37.776599 | 6 | |
| 4 | 39.223401 | 41.833333 | 44.443265 | 6 | |
| 3. (A,B combination) - A 95% CI on the treatment mean | | | | | |
| Level | Lower confidence limit | Mean | Upper confidence limit | Replicates | |
| A1B1 | 9.975666 | 13.666667 | 17.357668 | 3 | |
| A1B2 | 17.308999 | 21 | 24.691001 | 3 | |
| A1B3 | 26.975666 | 30.666667 | 34.357668 | 3 | |
| A1B4 | 32.642332 | 36.333333 | 40.024334 | 3 | |
| A2B1 | 21.308999 | 25 | 28.691001 | 3 | |
| A2B2 | 32.642332 | 36.333333 | 40.024334 | 3 | |
| A2B3 | 35.975666 | 39.666667 | 43.357668 | 3 | |
| A2B4 | 43.642332 | 47.333333 | 51.024334 | 3 | |

**Figure 7.12**   Two-way ANOVA table for ball throwing data, including ANOVA table to test main effects and interaction.

$25.417 - 37.083 = -11.666$. The mean distance of male students is longer than that of female students by 11.666 meters. The main effect of Training refers to the differences between four marginal means associated with the four training levels: 19.333, 28.667, 35.167, 44.443.

By not rejecting the third hypothesis (no interaction), we have a simpler prediction that states throwing distance can be computed by understanding the two factor effects separately. To get the effect of sex and training, we need only add the affects from each of the factors that are summarized in part 2. For that reason, the prediction model is sometimes called additive. If the interaction term was significant in the analysis, then every combination effect would be separately estimated using part 3. For example, the first effect in part 3 is labeled A1B1, which stands for a female (A1) who has no training (B1).

• • •

## 7.8   TWO-LEVEL FACTORIAL DESIGN ANALYSIS

A factorial design is a special term for multi-factor experiments where every factor level of every factor is observed so that main factor effects as well as interactions can be studied. Typically, the factors only have two possible levels, making it more reasonable to repeat the experiment at all the possible factor-level combinations. Example 7-7 can be viewed as a two-level factorial design because two input factors were used, and every combination of factor levels was utilized in the experiment.

Instead of labeling the factors as A1, A2, etc., eZ SPC assigns a letter (a,b,c) to each factor. Because each factor has just two possible levels, these levels are represented with a plus or minus. How the levels are assigned to the factors is completely arbitrary, but this way, any factor combination can be neatly described as a combination of the letters that represent the factor. For example, suppose we have four factors (a,b,c,d). There are $2 \times 2 \times 2 \times 2$ = 16 different factor level combinations (or treatments). If the experiment is run with factor a at the + level, b at the – level, c at the + level and d at the – level, the factor combination is symbolized using only the letters that are used at the plus level: ac. This is slightly awkward if the experiment is repeated with all factors set to the – level, since no letters are used. The result of running the experiment at all – levels is symbolized with (1). For a four factor experiment, the 16 combinations can be represented

$$(1), a, b, c, d, ab, ac, ad,$$

$$bc, bd, cd, abc, abd, bcd, acd, abcd.$$

Using a factorial design, we can compute the main effects of factors and interactions effects. The main effect is the difference between the average of high level or + level and the average of low level or – level. For example, the main effect of factor A is the difference between the average of A + level and the average of A – level.

## Example 7-7: Effect of lecture on student performance

To improve academic performance in undergraduate classes, a college decides to run an experiment to see if student performance is affected by lecture time length, the type of media used in lectures, and student/teacher interaction. These three factors are each considered at two different levels: lecture time per class (75 minutes versus 50 minutes), lecture materials (use of presentation slides versus use of blackboard and chalk), and lecture method (traditional lecture versus cooperative learning settings). Eventually, the experiment was tried for 32 different classes studying the same subject over four weeks, and student retention was based on the outcome of an identical test used in every class.

| Factor | Level | Description |
|--------|-------|-------------|
| A | – | 75 minute classes |
| A | + | 50 minute classes |
| B | – | lecture slides used |
| B | + | lecture on chalk board |
| C | – | traditional lecture |
| C | + | cooperative learning |

There are eight possible treatments from the combinations of levels of each factor and they are:

$$( \ (1), \ a, \ b, \ c, \ ab, \ ac, \ bc, \ abc \ ).$$

At level bc, for example, the students experience 75 minute classes (A–), lectures presented on a chalk board (B+) and cooperative learning methods are employed during lecture time (C+). Four students are assigned to each treatment combination so a total of 32 students are selected from the similar GPA groups. The number of replications is $n = 4$. This is a typical example of a $2 \times 2 \times 2$ ($2^3$) factorial design. The recorded scores of 32 students are as follows:

|  | 1 | 2 | 3 | 4 |
|-----|----|----|----|----|
| (1) | 73 | 78 | 72 | 80 |
| a | 70 | 79 | 75 | 77 |
| b | 77 | 81 | 80 | 82 |
| ab | 83 | 90 | 88 | 95 |
| c | 85 | 84 | 88 | 87 |
| ac | 74 | 85 | 91 | 90 |
| bc | 87 | 80 | 89 | 83 |
| abc | 89 | 94 | 93 | 98 |

For example, the first row (with the dot) represents the scores of the four students who had 75 minute classes (–), lecture slides (–) and a traditional lecture (–). The last line has the scores of the four students who had 50 minute classes (+), lecture on a chalk board (+) and experienced cooperative learning techniques in class (+). Because all the treatment levels are at the + levels, this combination is represented as abc. To input the factors in eZ SPC, go to the Design of Experiment submenu under the Analysis tab, and select Factorial Design. The program is set up to handle factors with two levels, like this one. Enter the number of factors (3) and number of replicates (4) for the experiment described in Example 7-7. Once the table is filled in, click on **Run Anova**. eZ SPC constructs a new window like Figure 7.13.

In Example 7-7, there are actually seven different hypotheses to be tested:

- Null Hypothesis 1: Factor A has no effect

| File(F) | Edit(E) | Tools(T) | Graph(G) | Analysis(A) | View(V) | Windows(W) | Help(H) |

| Experiment Condition | | | Treatment combination | Experiment Data | | | |
|---|---|---|---|---|---|---|---|
| A | B | C | | Experiment 1 | Experiment 2 | Experiment 3 | Experiment 4 |
| - | - | - | (1) | 73 | 78 | 72 | 80 |
| + | - | - | a | 70 | 79 | 75 | 77 |
| - | + | - | b | 77 | 81 | 80 | 82 |
| + | + | - | ab | 83 | 90 | 88 | 95 |
| - | - | + | c | 85 | 84 | 88 | 87 |
| + | - | + | ac | 74 | 85 | 91 | 90 |
| - | + | + | bc | 87 | 80 | 89 | 83 |
| + | + | + | abc | 89 | 94 | 93 | 98 |
| | | | | | | | Start ANOVA |

**Figure 7.13**    Input frame for factorial design used in Example 7-7: Effect of lecture on student performance.

- Null Hypothesis 2: Factor B has no effect

- Null Hypothesis 3: Factor C has no effect

- Null Hypothesis 4: A*B Interaction has no effect

- Null Hypothesis 5: A*C Interaction has no effect

- Null Hypothesis 6: B*C Interaction has no effect

- Null Hypothesis 7: A*B*C Interaction has no effect

The interactions from the fourth to seventh hypotheses are important. If any of them are significant, we will end up with a complicated relationship between input factors that will be difficult to explain. For example, if the B*C interaction is significant, then the effects of using slides (instead of chalk board) is different for classes that use cooperative learning compared to traditional learning. The three factor interaction A*B*C is especially hard to interpret; it essentially says that the A*B interaction actually depends on what level we have set for factor C. To be sure, we hope the three factor interaction is not significant. In other words, we really hope we do not reject Null Hypothesis 7. Any predictor with three two-factor interactions is pretty hard to interpret as well, but finding an interaction in an experiment like this can be a great learning experience. In any case, let's hope there are not too many two-way interactions that are considered significant.

The ANOVA for the three-factor experiment in Example 7-7 is shown in Figure 6-11. In the ANOVA table, Factors A, B, and C are significant at the level 0.05. That is, the p-values for each of these hypotheses are smaller than 0.05. The A*B interaction effect is significant at the significance level 0.01, but the other interactions are not significant at the significance level 0.05 (whew!). How do we interpret the result of our factorial experiment through the ANOVA table? Each of the individual factors is important in determining a student's test score. Perhaps more importantly, the effect of lecture time is different for classes that rely on slide presentation, compared to classes that rely on chalkboard presentation. Notice that we could interpret this as a B*A interaction instead of an A*B interaction if we want - the order of the factor is arbitrary. Although the relationship between the first two factors is nuanced, we can understand the effect of teaching method (traditional lecture versus cooperative learning methods) by looking solely at the main effect of factor C on the table.

· · ·

### 1. ANOVA Table

| Source | Sum of squares(SS) | DF | Mean square(MS) | F0 | p-value | Test result(significance level:0.05) |
|---|---|---|---|---|---|---|
| Factor A | 132.03125 | 1 | 132.03125 | 6.899837 | 0.014781051379689 | Factor A Reject H0 |
| Factor B | 318.78125 | 1 | 318.78125 | 16.659227 | 0.0004286742193 | Factor B Reject H0 |
| Factor C | 427.78125 | 1 | 427.78125 | 22.35547 | 0.00008289237272 | Factor C Reject H0 |
| A*B | 185.28125 | 1 | 185.28125 | 9.682635 | 0.004752225878482 | Reject H0 |
| A*C | 0.28125 | 1 | 0.28125 | 0.014698 | 0.904514394821221 | Fail to reject H0 |
| B*C | 57.78125 | 1 | 57.78125 | 3.019597 | 0.095081453367884 | Fail to reject H0 |
| A*B*C | 0.03125 | 1 | 0.03125 | 0.001633 | 0.968099286554434 | Fail to reject H0 |
| Error | 459.25 | 24 | 19.135417 | | | |
| Sum | 1581.21875 | 31 | | | | |

### 2. Factors Effect

| Factors | Factors Effect | Sum of squares(SS) | Percent contribution(%) |
|---|---|---|---|
| Factor A | 4.0625 | 132.0312 | 8.349967 |
| Factor B | 6.3125 | 318.7812 | 20.160477 |
| Factor C | 7.3125 | 427.7812 | 27.053894 |
| A*B | 4.8125 | 185.2812 | 11.717623 |
| A*C | -0.1875 | 0.2812 | 0.017787 |
| B*C | -2.6875 | 57.7812 | 3.654222 |
| A*B*C | 0.0625 | 0.0312 | 0.001976 |

**Figure 7.14**    ANOVA table for the data in Example 7-7.

## 7.9   WHAT DID WE LEARN?

- Simple linear regression can be used to construct a model that can predict process outputs as a function of process inputs.

- The coefficient of determination describes the proportion of variability in the output that can be explained by changing values of the input.

- The residual plot and normal probability plot are used to validate the results of the linear regression.

- A key objective of experimental design is to find input factors that matter to the output, and knowing what levels to set those inputs to achieve an output close to the target value of the process.

- In cases where the inputs do not determine the target mean, they can sometimes be set to values that minimize the variability of output.

- In experimental design, blocking is used to reduce the effect of variability that comes from differences in experimental units.

- When the level of one input factor matters in the way another input factor affects the output, these input factors are said to interact.

- A $2^k$ factorial design means we are using k input factors, each set at two different levels, thus there are $2^k$ different input combinations that can be used in the experiment.

- Analysis of variance is typically used to find out what input factors influence the process output, whereas regression is typically used to find out far that influence extends.

## 7.10   TEST YOUR KNOWLEDGE

1. In which of the following is not the principle of experimentation?

  a. Principle of normality

  b. Principle of blocking

  c. Principle of replication

  d. Principle of randomness

**2.** Which of the following correctly describes the coefficient of determination?

a. It is between $-1$ and 1.

b. It is the square of sample standard deviation.

c. It is between 0 and 1.

d. If it is 0.5, then 50% of data are used in the regression analysis.

**3.** Which of the following is not correct?

a. Negative value of $r$ means that two variables are negatively related.

b. Positive value of $r$ means that $Y$ value increases as $X$ value increases.

c. Correlation coefficient, r, must be between $-1$ and 1.

d. Correlation coefficient, r, must be between 0 and 1.

**4.** In an experiment, the engineer wants to study the relationship between 4 factors such as A, B, C, and D and all interactions. He is going to set two levels of each factor. Which experimental design is appropriate?

a. One way Design

b. Two way Design

c. Two level Factorial Design

d. Latin square design

**5.** In one-way analysis of variance, suppose the null hypothesis is rejected. Which of the following is not correct?

a. All treatment means are different.

b. Some treatment means are different.

c. We don't know if any treatment means are different.

d. All treatment means are equal.

**6.** In two-level factorial design experiments, five factors are considered for the study. How many treatments will be available?

a. 10

b. 25

c. 32

d. 64

7. In Example 7-7, what do the treatments b and c have in common?

    a. 75 minute classes

    b. 50 minute classes

    c. lecture slides used

    d. lecture on chalk board

8. Which of the following is not the treatment of the two level factorial design with three factors (A,C,E)?

    a. a

    b. ce

    c. ace

    d. d

9. Which of the following is correctly defines the main effect of A?

    a. It is the difference of + level of A and − level of A.

    b. It is the variance of factor A.

    c. It is the difference of average of high level of A and average of low level of A.

    d. It is the average of + level of A and − level of A.

10. Which of the following is not included in the experimentation steps?

    a. Select experimental units.

    b. Select factors.

    c. Select the experimental design.

    d. Select experimenter.

## EXERCISES

**7.1**   A company's human resources department studied the relationship between the number of years an employee has worked at the company and their level of loyalty (rated 1 for lowest to 5 for highest) to the company. They randomly selected 15 employees to survey, and the results are listed in the file Loyalty.ezs. What can you say about the relationship between an employee's work period and the worker's loyalty level?

**7.2**   A meat packing company wants to learn more about the relationship between quality-related training expenditures and defective rate in their packaged meats. A lag time of one month will be used because training effects are expected to lag behind actual training. Data are collected for a year period and given in Training.ezs. Expenditures are in listed in US dollars.

Use a simple linear regression analysis to determine the relationship between training expenditures and quality level.

**7.3** Relative humidity is essential component of a comfortable home environment. An experimental study was done to find the most comfortable relative humidity level for home living. It appears that exposure to dust mites may cause asthma and allergies, and dust mite control could be related to humidity levels. The experimenter measured the number of dust mites in 1.0 cubic feet at the four houses in the study at five selected relative humidity levels (40%, 45%, 50%, 55%, 60%). The number of dust mites found in one sampled cubic foot (for each house and humidity level combination) are given in the table below. Analyze the data and draw conclusions.

| % Humidity | House 1 | House 2 | House 3 | House 4 |
|:----------:|:-------:|:-------:|:-------:|:-------:|
| 40 | 0 | 2 | 0 | 1 |
| 45 | 2 | 3 | 4 | 2 |
| 50 | 10 | 12 | 14 | 13 |
| 55 | 32 | 29 | 43 | 48 |
| 60 | 102 | 114 | 129 | 119 |

**7.4** Students are competing to make a rubber band glider for longer time in air. They use same rubber band gliders with three rubber bands. Students were instructed to consider two factors for the study: the number of rubber band rotations (100, 200, 300) and the releasing angle of the glider (15°, 30°). They recorded the time (in seconds) from the release to landing on the ground. The data are given in the file `Glider.ezs`. Analyze the data and draw conclusions.

**7.5** A factorial design is used to investigate the effects of four factors on the indoor air quality for an office. The factors are A = relative humidity (45%, 55%), B = temperature (68, 78 degrees Fahrenheit), C = ventilation rate (10, 20 cubic feet per minute per person), and D = carbon dioxide (500, 1000 ppm). In total, there are $2 \times 2 \times 2 \times 2 = 16$ different factor level combinations to consider. At each level, employees scored the "comfortableness" of the indoor air quality based on a scale of 1 (lowest) to 10 (highest). The data are found in `AirQuality.ezs`. Analyze the data and test which factors are important to determining indoor air quality.

# CHAPTER 8

# END MATERIAL

In this section we feature a final exam consisting of 100 multiple choice questions along with the answers. The questions are basic, but they comprehensively span the entire text up to this point. If you have read the whole book, you should be able to finish this exam in two hours. Grade yourself on the following curve:

**A** If your score is between 90 and 100. This is an excellent score and shows that you have mastered the material from all seven chapters. Well done!

**B** If you scored between 75 and 89 give yourself a B. This is a pretty good score, but you might want to look over the problems you checked wrong and find out if there are specific areas that require more study.

**C** If you scored between 60 and 74, you get a grade of C. This signifies that you understood most of the basic material from the text, but you certainly could improve your mastery of these basic tools and techniques. Check to find out if your scores were especially low on any particular subject.

**F** If you scored fewer than 60 points on the exam, it shows that you have not absorbed the ideas or understood the tools from Chapters 2 through 7. You might want to consider reading the material more carefully and if you have not completed all of the exercises, participate in those exercises more the next time through. For math-phobic readers, it sometimes takes more than one try at this kind of material, so don't feel bad. We really hope you try it again.

## 8.1   FINAL EXAM

**100 Final Exam Questions:** Multiple choice answers with only one correct answer to each question. Solutions are found at the end of the exam.

**1.** In addition to its Coupe 2.0 sports car, the Hyundai motors company also sells a limited edition of the car that features different colors, improved wheels, and a leather upholstered interior. The differences between the standard Hyundai Coupe 2.0 and the special edition represent differences in

   a. Quality.

   b. Grade.

   c. Quality and Grade

   d. Neither Quality nor Grade.

**2.** In an agricultural process of growing and harvesting soybeans, which of the following represent chance variation in the primary output?

   a. Amount of pesticides used during growing season

   b. Number of treatments applied to soil.

   c. Brand and model of harvester used in bringing in the crop.

   d. Amount of rain water obtained during growing season

**3.** A trucking company requires each of its trucks receive an oil change for every 5,000 miles driven. What kind of quality cost does this entail?

   a. Prevention cost

   b. Appraisal cost

   c. Internal Failure Cost

   d. External Failure Cost

**4.** Suppose that the Bianchi bicycle company in Italy loses 30% of its business after it ended the use of their trademark turquoise (called "Bianchi Green" or "Celeste") color for their new model bikes. What quality dimension of the process affected the sale of bikes?

   a. Aesthetics
   b. Durability
   c. Performance
   d. Serviceability

**5.** Which of the following statistics are reported in a box plot?

   a. The X-MR control chart
   b. The u control chart
   c. The sample variance
   d. The quartiles of the data

**6.** A manufacturing company has identified 20 sources of error that cause defects in their product, and it turns out that the four major sources are attributed to 80% of the defects. This is an example of

   a. Lean Six Sigma
   b. Skewness
   c. The Pareto principle
   d. A frequency histogram

**7.** If the increase in one input variable (X) causes the increase of the output variable (Y), these variables are

   a. Independent
   b. Positively correlated
   c. Negatively correlated
   d. Explained graphically using a scatter plot.

**8.** In repeated samples of size n = 30, the sample average is calculated and stored. If we construct a histogram of the sample averages, the shape of the distribution will be bell-shaped. This is due to

   a. The Pareto Principle
   b. The Central Limit Theorem
   c. The p-value
   d. Type I Error in Tests of Hypotheses

**9.** If your hypothesis test produces a test statistic with a p-value of 0.1027, which of the following actions are appropriate?

   a. If the Type I error is set to 0.05, we should reject the null hypothesis.

   b. If the Type I error is set to 0.05, we should not reject the null hypothesis.

   c. No matter what Type I error is considered, we should reject the null hypothesis

   d. Even if we know the Type I error rate, there is not enough information to decide whether or not to reject the null hypothesis.

**10.** We want to show that golf ball A will go further golf ball T when they are hit by an average golfer. The average driving distance obtained by ball A is $A_0$ meters, and average driving distance obtained by ball T is $T_0$ meters. If we experiment by having 100 average golfers hit each of the golf balls and measure the driving distance, which of the following hypothesis is appropriate for this test?

   a. $H_0 : A_0 = T_0$ versus $H_1 : A_0 \neq T_0$

   b. $H_0 : A_0 \leq T_0$ versus $H_1 : A_0 > T_0$

   c. $H_0 : A_0 \geq T_0$ versus $H_1 : A_0 < T_0$

   d. $H_0 : A_0 \neq T_0$ versus $H_1 : A_0 = T_0$

**11.** A river is being monitored by measuring the amount of contaminants found near an effluent release pipe. Two samples (X, Y) are measured and recorded along with two older measurements (A, B) from a week earlier. The sample mean is (X+Y+A+B)/4. To emphasize the more recent measurements, an alternative estimator for effluent release is calculated as (X+Y)/3 + (A+B)/6. This represents

   a. A sample mean

   b. A sample median

   c. A weighted mean

   d. A sample mode

**12.** Which of the following graphical methods can be used to monitor a system based on counting the number of defects it produces?

   a. CUSUM control chart.

   b. $\bar{x}$- R control chart.

   c. Histogram

   d. p control chart.

**13.** If a manufacturing process produces 20% defects, what is its average run length $(ARL_0)$?

  a. 5

  b. 20

  c. 100

  d. 370

**14.** A machine fills bags with sugar based on a target amount of 2 lbs. Once per hour, 12 bags are weighed to see if the machine is filling the bags close to this target value. Using these data, what graphical procedure should be used to monitor this machine?

  a. $\bar{x}$ - R control chart

  b. $\bar{x}$ - s control chart

  c. p control chart

  d. x - MR control chart

**15.** In the last problem, if only one bag of sugar can be sampled and weighed per sampling interval, what graphical procedure should be used to monitor this machine?

  a. $\bar{x}$ - R control chart

  b. $\bar{x}$ - s control chart

  c. p control chart

  d. x - MR control chart

**16.** In which of the following control charts can the upper and lower specification limits change during the monitoring process?

  a. $\bar{x}$ - R control chart

  b. $\bar{x}$ - s control chart

  c. p control chart

  d. x - MR control chart

**17.** A product that meets the required specifications of the manufacturer is said to be

  a. Conforming

  b. Non Conforming

  c. Confounding

  d. Stable

**18.** To compare the products of two manufacturing companies using sample product measurements from both, we should use

a. $\bar{x}$ - R control chart

b. $\bar{x}$ - s control chart

c. Two sample t-test

d. Goodness of Fit test

**19.** If the process capability index is Cp = 4.00, what proportion of the specification band is used up by the measured output?

a. 25.00 %

b. 50.00 %

c. 96.00 %

d. 99.60 %

**20.** In testing to see if the distribution of the data is normal, suppose the normal probability plot reveals a noticeably curved pattern in the plotted points. Which of the following statements are true?

a. The observed data have a distribution with a bell-shaped curve.

b. The sample range should be used instead of the standard deviation.

c. The observed data have a Poisson distribution.

d. The process capability analysis can lead to erroneous results.

**21.** When monitoring a process, a CUSUM control chart is preferred over an EWMA control chart if

a. We want to give more weight to recent measurements.

b. We want to give less weight to recent measurements.

c. We want to give the same weight to all measurements.

d. the measurements fall outside the decision interval.

**22.** Measurement systems analysis can not be used to assess which of the following?

a. The bias inherent in the measurements.

b. The gages that measure outputs.

c. Natural variation in the system.

d. The capability index.

**23.** An experimenter wants to characterize the relationship between four input factors. Learning about potential factor interaction is also considered important. Due to limited resources, only two levels are used for each of the two factors. Which experimental design is appropriate?

   a. One way Design

   b. Two way Design

   c. Two level Factorial Design

   d. Latin square design

**24.** In two-level factorial design experiment, four different input factors are considered for the study. How many treatment combinations will be available?

   a. 16

   b. 32

   c. 64

   d. 1024

**25.** In a linear regression, the coefficient of determination is computed to be 0.50. Which of the following choices correctly interprets this value?

   a. 50% of regressions made this way are significant

   b. The type I error is 0.50

   c. 50% of the response variation is explained by the predictor.

   d. the coefficient of determination has no direct interpretation.

**26.** The graphical tool that lists potential error sources on lines which branch off from their effects is called a

   a. Histogram

   b. Factorial Design

   c. Branch Algorithm

   d. Cause and Effect Diagram

**27.** To increase the average run length of a process that is currently stable, which of the following techniques would be effective?

   a. Monitor the process with a CUSUM chart

   b. Increase the upper control limit

   c. Compute the process capability index

   d. Monitor the process with a CV chart

[28 - 44] A pharmaceutical company manufactures a liquid drug and fills the product into a plastic bag. The Form/Fill/Seal (FFS) process includes forming a bag, filling the bag with the drug, and sealing the bag. The data from this process are collected for 20 days and are listed in the file LiquidDrug.ezs. Operators found that there are many empty bags produced. They established a quality control (QC) circle to solve the empty bag production problem.

**28.** The QC team calculated the costs of labor, raw material, parts, and general expenses of this process and analyzed which portion of cost is the most. Which one of the following graphs is appropriate?

   a. Pie chart

   b. Box plot

   c. Bar graph

   d. Line graph

**29.** They collected the daily number of empty bags data for the last 20 days (see LiquidDrug.ezs) and wanted to check the trend. Which one of the following graphs is appropriate?

   a. Pie chart

   b. Normal probability plot

   c. Radar chart

   d. Line graph

**30.** They calculated the defective rate and the ratio of the daily number of empty bags over the daily number of bags that were produced. To ascertain whether the process is in statistical control, which of the following control charts is appropriate?

   a. np control chart

   b. c control chart

   c. p control chart

   d. u control chart

**31.** The team discussed the causes of empty bags and grouped those causes into a few related categories: machine, material, people, method. What kind of QC tool should be used with this information?

   a. Check sheet

   b. Pareto diagram

   c. Fishbone chart

   d. Flow chart

**32.** Which of the following is the sample mean for the daily number of empty bags?

   a. 620.55

   b. 630.44

   c. 632.66

   d. 640.55

**33.** What is the sample standard deviation for the daily number of empty bags?

   a. 0.000

   b. 0.597

   c. 23.13

   d. 24.44

**34.** Which of the following is the sample median for the daily number of empty bags?

   a. 644.5

   b. 640.5

   c. 642

   d. 650.5

**35.** Which of the following is the range for the daily number of empty bags?

   a. 82

   b. 83

   c. 84

   d. 85

**36.** The company wanted to see if there is much variability in the daily number of empty bags that are produced. Which graph is appropriate?

   a. Box plot

   b. Scatter plot

   c. u control chart

   d. Line graph

**37.** To understand the relationship between the daily number of bags produced (X) and the daily number of empty bags produced (Y), what kind of QC tool should be used?

   a. Scatter plot

   b. Normal probability plot

   c. $\bar{x}$ - R control chart

   d. Bar graph

**38.** Which of the following is the correlation coefficient value?

   a. 0.34

   b. −0.40

   c. −0.55

   d. 1.21

**39.** Which of the following represents a 95% confidence interval for the mean daily number of empty bags?

   a. (631.10, 650.00)

   b. (629.11, 651.99)

   c. (624.91, 656.19)

   d. (623.90, 657.17)

[40 - 42] The process engineer claims that the process defective rate is less than 3%. The QC team randomly sampled 150 bags and found 6 empty bags. The team wants to test the engineer's claim, using 0.05 as the level of significance.

**40.** Which of the following is correct alternative hypothesis?

   a. $p > 0.03$

   b. $p \neq 0.03$

   c. $p = 0.04$

   d. $p < 0.03$

**41.** Which of the following is the correct z-value for the test?

   a. 0.718

   b. 0.123

   c. 1.201

   d. 0.826

**42.** Which of the following is the correct p-value for the test?

a. 0.05

b. 0.01

c. 0.764

d. 0.1

[**43 - 44**] The company formally investigated the different types of noncon-formities in the product: no opening, port error, no port, printing error, bag overlapping, and others, which result into empty bag. Based on the data from LiquidDrug.ezs, the types of nonconformity are given in the table below.

| Nonconformity | Number |
|---|---|
| Empty Bags | 344 |
| No Opening | 110 |
| Port error | 110 |
| No Port | 90 |
| Over- sealing | 62 |
| Printing error | 19 |
| Bag Overlapping | 17 |
| Other | 19 |

**43.** Which of the QC tools should be used if the company wants to know which nonconformity is the most important?

a. Cause and effect diagram

b. Pareto diagram

c. Radar chart

d. Bar graph

**44.** From the table above, what is the cumulative percentage for no opening and port error?

a. 68.684%

b. 45.4%

c. 34.4%

d. 66.1%

[45 - 59] The Bridge Maneuvering Systems (BMS) are standard control systems for commercial shipping. A shipbuilding company is attempting to reduce the BMS manufacturing time because the number of BMS orders is rapidly increasing. In order to reduce the manufacturing time, engineers formed a task force team (TFT), and they looked into the manufacturing processes and measured the standard process time in man-hour units:

| Assembly Preparation | Assembly | Test | Finishing/Shipping |
|:---:|:---:|:---:|:---:|
| 5.1 | 17.5 | 4.1 | 1.3 |

This shipbuilding company collected manufacturing time data for 10 manufactured BMS sets. The data are listed in man-hour units below:

| No. | 1 | 2 | 3 | 4 | 5 | 6 | 7 | 8 | 9 | 10 |
|---|---|---|---|---|---|---|---|---|---|---|
| Time | 28.4 | 32.6 | 31.3 | 32.2 | 36.5 | 33.6 | 30.7 | 38.2 | 30.7 | 34.2 |

**45.** What is the appropriate tool for identifying the time proportion of each process?

   a. Line graph

   b. Radar chart

   c. Pie chart

   d. Bar graph

**46.** What is the appropriate tool for checking the trend in manufacturing time?

   a. Bar graph

   b. Line graph

   c. Control chart

   d. Scatter plot

**47.** What is the appropriate tool for checking the variation in manufacturing time?

   a. Bar graph

   b. Box plot

   c. Line graph

   d. Control chart

**48.** Which of the following represents the sample mean manufacturing time?

   a. 32.84

   b. 33.47

   c. 32.18

   d. 34.02

**49.** Which of the following represents the sample variance of manufacturing time?

   a. 8.47

   b. 8.10

   c. 9.22

   d. 7.89

**50.** Which of the following is appropriate for checking the normality in the distribution of manufacturing time?

   a. Residual plot

   b. Box plot

   c. Scatter plot

   d. Normal probability plot

**51.** Which of the following intervals represents a 95% confidence interval for the mean manufacturing time?

   a. (31.15, 34.53)

   b. (30.76, 34.92)

   c. (29.85, 35.83)

   d. (27.35, 37.18)

[52 - 54] The TFT decided to test whether the mean BMS manufacturing time ($\mu$) exceeds the standard manufacturing time, which is 28 man-hours. Test this hypothesis using 0.01 as the test's level of significance.

**52.** Which of the following is correct alternative hypothesis?

   a. $\mu = 28$

   b. $\mu < 28$

   c. $\mu > 28$

   d. $\mu \neq 28$

**53.** Which of the following is the correct t-value test statistic for the hypothesis test?

  a. 2.33

  b. 3.41

  c. 5.26

  d. 7.48

**54.** Corresponding to the test outcome, which of the following is the corrective p-value?

  a. 0.01

  b. 0.00026

  c. 0.0001

  d. 0.0005

[55 - 58] Next, the TFT studied potential reasons for the delayed manufacturing time and found that their operators do not use the SOP (Standard Operating Procedure). After training all of the operators using the SOP, they measured the production time for the same BMS model. The data are given in the table below.

| Number | 1 | 2 | 3 | 4 | 5 | 6 | 7 | 8 | 9 | 10 |
|--------|------|------|------|------|------|------|------|------|------|------|
| **Before** | 29.1 | 31.4 | 32.5 | 31.9 | 30.6 | 29.9 | 32.4 | 31.7 | 30.8 | 32.9 |
| **After** | 27.2 | 26.9 | 28.8 | 27.6 | 28.4 | 28.1 | 27.9 | 28.1 | 28.0 | 28.3 |

**55.** Which of the following numbers represents mean difference time between before and after the SOP training?

  a. 3.39

  b. 3.21

  c. 3.48

  d. 3.76

**56.** Which of the following is correct sample standard deviation of this difference time?

  a. 1.126

  b. 1.382

  c. 1.114

  d. 1.982

**57.** Which of the following represents the 95% confidence interval for mean difference time?

   a. (2.737, 4.043)

   b. (2.585, 4.195)

   c. (2.233, 4.547)

   d. (2.112, 4.672)

**58.** If the TFT tests the null hypothesis $H_0$: the mean difference equals 2 vs $H_1$: the mean difference is greater than 2, which of the following numbers is the correct p-value?

   a. 0.0018

   b. 0.1

   c. 0.01

   d. 0.05

**59.** If the manufacturing time is effectively reduced by training operators, which of the following graphical procedures should be used to show the reduction in mean time?

   a. Two histograms

   b. Two box plots

   c. Two line graphs

   d. Two bar graphs

[60 - 74] In the automobile door assembly process, the door gap measurement represents an important quality characteristic. The door gap specification limits are $4.5 \pm 0.5$ (mm), and any autos with door gap measurements outside this specified range are non conforming. The door gap measurement data are found in the file DoorGap.ezs.

**60.** Which graphical procedure is best suited to illustrate the distribution of the data?

   a. Histogram

   b. Bar graph

   c. Pareto diagram

   d. Line graph

**61.** Which of the following charts is appropriate for monitoring the process?

   a. p control chart

   b. x - MR control chart

   c. $\bar{x}$ - R control chart

   d. $\bar{x}$ - s control chart

**62.** Which of the following assumptions is not required for calculating the process capability index?

   a. The quality characteristic is normally distributed.

   b. The process is in statistical control.

   c. The process mean is centered between the two sided specification limits.

   d. The process variance must be minimized.

**63.** Which of the following are the correct specification limits for the door gap?

   a. (4.0, 5.0)

   b. (4.45, 4.55)

   c. (4.5, 5.0)

   d. (4.5, 4.5)

**64.** Which of the following numbers is the centerline of $\bar{x}$ -R control chart?

   a. 4.265

   b. 4.500

   c. 4.555

   d. 4.562

**65.** Which of the following values is the UCL of the control chart?

   a. 4.778

   b. 4.812

   c. 4.899

   d. 4.998

**66.** Which of the following values is the LCL of the control chart?

   a. 4.225

   b. 4.345

   c. 4.123

   d. 4.453

**67.** Which of the following statements represents the correct definition of the warning limits in the control chart? "The are computed at..."

    a. One standard deviation below and above the centerline.

    b. Two standard deviations below and above the centerline.

    c. Three standard deviations below and above the centerline.

    d. Four standard deviations below and above the centerline.

**68.** Which of the following is the process capability index value?

    a. 0.500

    b. 0.670

    c. 1.333

    d. 1.678

**69.** Which of the following is the mean of ranges?

    a. 0.584

    b. 0.552

    c. 0.548

    d. 0.500

**70.** Which of the following is the UCL of the R control chart?

    a. 1.100

    b. 1.235

    c. 1.255

    d. 1.300

**71.** Which of the following is the LCL of the R control chart?

    a. 0.00

    b. 1.00

    c. 0.50

    d. 0.75

**72.** If the Cp value is less than 1.0, which of the following represents an acceptable diagnosis of the process?

    a. The process is excellent.

    b. The process needs to be improved.

    c. The process must be monitored by $\bar{x}$ - s control chart.

    d. We can't tell about this process.

**73.** Which of the following statements is not correct?

    a. We can apply the $\bar{x}$ - R control chart in any process.

    b. When the process mean shift is small, we apply a CUSUM control chart.

    c. When the process mean shift is small, we apply an EWMA control chart

    d. When the process mean shift is large, we apply $\bar{x}$ - R control chart.

**74.** When the variance is a function of the mean, the coefficient of variation is an appropriate measure for process variability. Which of the following control charts is appropriate in this case?

    a. EWMA control chart

    b. Shewhart control chart

    c. CV control chart

    d. CUSUM control chart

[75 - 85] In a Beijing school, a physical examination was conducted to check students' health condition. Ten 18-year-old female students were randomly selected and their heights and weights were recorded. The data are given in the file HeightWeight.ezs, where Y = weight (kg) and X = height (cm) are measured for n = 10 eighteen-year-old female students.

**75.** In order to illustrate the relationship between height and weight, which of the following graphical tools is appropriate?

    a. Residual plot

    b. Box plot

    c. Scatter plot

    d. Normal probability plot

**76.** From your graphical analysis of the data, which of the following is correct in regard to the relationship between weight and height for the female students?

    a. Independent

    b. Positively correlated

    c. Negatively correlated

    d. Uncorrelated

**77.** Which of the following methods is the appropriate means to estimating the regression line slope and intercept?

a. Pareto method

b. Least squares method

c. Central Limit theorem

d. Normality test

**78.** If we use Y = weight for the response and X = height as the predictor, what is the estimated value for the slope of the regression line?

a. 0.733

b. 0.838

c. 1.000

d. 1.375

**79.** Which of the following is correct for the coefficient of determination?

a. 0.702

b. 0.838

c. 0.931

d. 0.937

**80.** If we want to check to find out if the variance of error is constant, which of the following is appropriate?

a. Normal probability plot

b. Scatter plot

c. Residual plot

d. Histogram

**81.** If a new eighteen-year-old female student joins the class, and her height is 166 cm, which of the following is the predicted value for her weight?

a. 50.8 kg

b. 53.2 kg

c. 55.6 kg

d. 58.2 kg

**[82 - 84]** An electronics company manufactures power supplies for a personal computer. They produce 500 power supplies during each shift, and each unit is subjected to a 24-hour burn-in period. The number of units failing during each shift of a 24-hour test is listed for 150 consecutive shifts in the file PowerSupply.ezs.

**82.** Based on an np control chart, how many times was this process out of control?

   a. 0

   b. 1

   c. 2

   d. 3

**83.** To decide whether the process was out of control, what was the calculated upper control limit for the chart?

   a. 1.350

   b. 2.700

   c. 6.933

   d. 14.78

**84.** The maximum number of failures in any shift was 17, and the minimum was 1, so the range is 16. What is the interquartile range (IQR) for the failure number?

   a. 4.00

   b. 5.25

   c. 8.00

   d. 8.13

[85 - 86] The data in `PhoneCable.ezs` represent the number of nonconformities per 1000 meters in telephone cable computed from 20 independent samples. From analysis of these data, would you conclude that the process is in statistical control?

**85.** Which of the following charts is appropriate for analyzing the phone cable data?

   a. p control chart

   b. c control chart

   c. $\bar{x}$ - R control chart

   d. $\bar{x}$ - s control chart

**86.** Based on your chosen control chart, how many times was this process out of control?

   a. 0

   b. 1

   c. 2

   d. 3

[87 - 94] An engineer is interested in the effects of cutting speed (A) and cutting angle (B) on the flatness of the surface of pulley that is being manufactured for special use in an automobile. Flatter is better, so the target is 0.00, and the upper specification limit of flatness is 0.05. A 2x2 set of factorial experiments with 3 replicates was conducted and the data are given in the table below.

| Treatment | Replicates | | |
|-----------|-------|-------|-------|
| (1) | 0.032 | 0.039 | 0.034 |
| a | 0.026 | 0.026 | 0.028 |
| b | 0.034 | 0.034 | 0.035 |
| ab | 0.040 | 0.047 | 0.046 |

**87.** Which of the following statements about factorial experiments is true?

a. We can always ignore factor interactions

b. We can estimate all main effects and interaction effects

c. One-factor-at-a-time approach is better a factorial design.

d. Best guess approach is better than a factorial design.

**88.** Which of the following is the estimated main effect of cutting speed?

a. 0.00083

b. 0.03467

c. 0.03550

d. 0.03933

**89.** Which of the following is the estimated main effect of cutting angle?

a. 0.00043

b. 0.00085

c. 0.00430

d. 0.00850

**90.** Which of the following is the estimated effect of interaction between cutting speed and cutting angle?

a. 0.009

b. 0.018

c. 0.092

d. 0.046

**91.** Which of the following pieces of information is not listed in the analysis of variance (ANOVA) table?

    a. Sum of squares

    b. Degrees of freedom

    c. Main effects

    d. Mean squares

**92.** Which of the following is correct for the error degrees of freedom in this experiment?

    a. 11

    b. 10

    c. 9

    d. 8

**93.** The p-value of the test for $H_0$: Factor B is not effective versus $H_1$: Factor B is effective is p = 0.0005. Which of the following is the correct conclusion if the level of significance set for this test is 0.01?

    a. Reject H1.

    b. Fail to reject Ho.

    c. Reject Ho.

    d. We can't tell due to lack of information.

**94.** Which is not an experimental principle used in the design of this experiment?

    a. Principle of exactness

    b. Principle of randomness

    c. Principle of blocking

    d. Principle of replication

[95 - 100] Some quality characteristics for a particular manufactured automobile part include run out, gap difference, flatness, and concavity. The measurements of 100 parts for each quality characteristic and its specifications are given in the file AutoParts.ezs.

**95.** Which of the two quality characteristics are most highly correlated

    a. Run Out and Flatness

    b. Run Out and Concavity

    c. Gap Difference and Concavity

    d. Flatness and Concavity

**96.** Construct an EWMA control chart for the Run Out characteristic using 0.09 as a target value. How many times, out of 100, was the process out of control?

    a. 0
    b. 1
    c. 2
    d. 3

**97.** For the concavity characteristic, construct a process capability analysis using lower and upper tolerances of 0.01 and 0.03. What is the computed Cp statistic?

    a. 0.779
    b. 0.977
    c. 1.000
    d. 1.333

**98.** In the normal probability plot, eZ-SPC also computes the Shapiro-Wilk test statistic. If this number is close to one, the data are presumed to have a normal distribution. Otherwise, the statistic will be closer to zero. According to the Shapiro-Wilk statistic, which quality characteristic is least likely to appear normally distributed?

    a. Run Out
    b. Gap Difference
    c. Flatness
    d. Concavity

**99.** Produce a box plot for flatness and concavity (select both columns to ensure the chart contains both box plots). Which of the following statements is proved by the graph?

    a. The mean for concavity and flatness are the same
    b. Flatness and concavity are highly correlated
    c. The median of flatness data is larger than the upper quartile of the concavity data
    d. Both flatness and concavity are distributed normally

**100.** For the normal probability plot of the flatness measurements, which of the following statements is true?

    a. A curved line means the measurements are attribute data
    b. The Shapiro Wilk statistic is 0.3109
    c. Plot is generated by central limit theorem
    d. A straight line indicates normality

## 8.2   FINAL EXAM SOLUTIONS

This page contains the multiple-choice answers to the 100 Final Exam questions from the previous pages.

| | | | | | | | |
|---|---|---|---|---|---|---|---|
| 1. | c | 26. | d | 51. | b | 76. | b |
| 2. | d | 27. | b | 52. | c | 77. | b |
| 3. | a | 28. | a | 53. | c | 78. | a |
| 4. | a | 29. | d | 54. | b | 79. | a |
| 5. | d | 30. | c | 55. | a | 80. | c |
| 6. | c | 31. | c | 56. | a | 81. | b |
| 7. | b | 32. | d | 57. | b | 82. | d |
| 8. | b | 33. | d | 58. | a | 83. | d |
| 9. | a | 34. | a | 59. | d | 84. | b |
| 10. | b | 35. | c | 60. | a | 85. | b |
| 11. | c | 36. | a | 61. | c | 86. | d |
| 12. | d | 37. | a | 62. | d | 87. | b |
| 13. | a | 38. | b | 63. | a | 88. | a |
| 14. | b | 39. | b | 64. | d | 89. | d |
| 15. | d | 40. | d | 65. | c | 90. | c |
| 16. | c | 41. | a | 66. | a | 91. | c |
| 17. | a | 42. | c | 67. | b | 92. | d |
| 18. | c | 43. | b | 68. | b | 93. | c |
| 19. | a | 44. | a | 69. | a | 94. | a |
| 20. | d | 45. | c | 70. | b | 95. | d |
| 21. | b | 46. | b | 71. | a | 96. | a |
| 22. | d | 47. | b | 72. | b | 97. | b |
| 23. | c | 48. | a | 73. | a | 98. | d |
| 24. | a | 49. | a | 74. | c | 99. | c |
| 25. | c | 50. | d | 75. | c | 100. | d |

## 8.3   TEST YOUR KNOWLEDGE: ANSWERS

At the end of Chapters 2 through 7, there is a section titled

### TEST YOUR KNOWLEDGE.

In each of these chapters, this section contains ten multiple choice questions with four possible answers (a, b, c, d). In some cases, more than on answer can be the correct one.

Check your answers with these answers below. If you achieved eight or more correct answers in a given quiz, you have shown a mastery of the material from that chapter. If you scored six or fewer correct answers on a quiz, you need to go back to review the material before proceeding to the next chapter.

|  | 1. | 2. | 3. | 4. | 5. | 6. | 7. | 8. | 9. | 10 |
|---|---|---|---|---|---|---|---|---|---|---|
| Chapter 1 | d | b | b | a | c | a | c | d | a | b |
| Chapter 2 | a,d | c | b,d | b | d | b,c | c | a,d | a,c | c,d |
| Chapter 3 | a,c | a,d | b | a | b | d | a | c | a | c |
| Chapter 4 | a | a | b | a | b | d | b | c | a | b |
| Chapter 5 | d | d | c,d | b | d | b | a | d | c | d |
| Chapter 6 | a | b | d | a | d | c | c | c | b | d |
| Chapter 7 | a | c | d | c | d | c | c | d | a | d |

# REFERENCES

Box, G. E. P., Hunter, W. G., and Hunter, J. S. (1978) *Statistics for Experimenters*, John Wiley & Sons, New York.

Deming, W. E. (1986) *Out of Crisis*, MIT Press, Cambridge, MA.

Garvin, D.A. (1987) "Competing on the Eight Dimensions of Quality", *Harvard Business Review*, 65, No. 6.

Garvin, D. A. (1988) *Managing Quality*. Free Press.

Ishikawa, K. (1985) *What is Total Quality Control? The Japanese Way*, Prentice-Hall, Englewood Cliffs, NJ.

Kang, C. W., Lee, M. S., Seong, Y. J., Hawkins, D. M., (2007) "A Control Chart for the Coefficient of Variation", *Journal of Quality Technology*, 39, pp. 151–158.

Montgomery, D.C., (2008) *Introduction to Statistical Quality Control*, 6th ed., John Wiley & Sons, New York.

Quesenberry, C. P(1997), *SPC Methods for Quality Improvement*, John Wiley & Sons, New York.

Shewhart, W. A. (1931) *The Control of Quality of Manufactured Products*, D. Van Nostrand, New York.

Stoumbos, Z. G., Reynolds, M. R., Ryan, T., Woodall, W. H. (2000) "The State of Statistical Process Control as We Proceed into the 21st Century", *Journal of the American Statistical Association*, 95, 992 – 996.

Webber, L. and Wallace, M. (2007) *Quality Control for Dummies*, Wiley Publishing, Inc., Hoboken, NJ.

# GLOSSARY

**8 Quality Dimensions.** These eight aspects of quality represent distinct properties for manufactured products: performance, features, reliability, conformance, durability, serviceability, aesthetics, and perceived quality. Conformance tells us if the product meets the specifications given by the manufacturer. Performance pertains to the displayed ability of the item's primary functions, unlike features, which provide secondary functions that the buyer may want from the product. Reliability means the product does not fail during its intended use period, while durability pertains to how long that period should be. Serviceability deals not only with the waiting time to repair or replace the product (once broken) but also refers to the nature of the company's interaction with the buyer. Aesthetics refer to the elegance, beauty or general attractiveness of the product.

**80-20 Rule** - see Pareto Principle.

**Additive.** In an Analysis of Variance, suppose the response is a function of two or more factors, called Factor A and Factor B. Suppose A and B have two input levels (low and high). Yield will change according to how Factors A and B are set at these two levels. If the affect of Factor A on yield does not depend on knowing what level is set for Factor B, then these factors are said to be additive. If the factors are not additive, they are said to interact.

**Aesthetics -** see 8 Quality Dimensions.

**Alternative Hypothesis.** Experimenters can state scientific claims by partitioning the possible experimental results into two hypotheses, the null and alternative. When claims are made about experimental outcomes, the alternative hypothesis usually coincides with the claim of some specific effect or an assignable cause. This claim is the complement of the null hypothesis, which usually implies the claimed effect is due to randomness or noise.

**Analysis of Variance.** Statistical procedures get this name by partitioning the observed variance of an experimental outcome according to explanatory variables. In the simplest example, different values of a single explanatory variable might lead to different expected outcomes. In this case the variance in the outcome is partitioned into two parts: variability due to differences in values of the explanatory variables and variability due to noise or measurement error in the outcome. When two or more explanatory variables are used to describe the outcome, the variance of the experimental result can be further partitioned.

**ANOVA -** see Analysis of Variance.

**Appraisal Cost -** see Quality Costs.

**Assignable Cause Variation.** The variation of product output due to unanticipated causes that need to be addressed in order to make the process efficient. Assignable causes consist of input errors, process degradation, faulty machinery, poor quality material input, or sloppy operation. Assignable cause variation differs from noise (or common-cause variation), which is due to natural variation in the output.

**Attribute Control Chart.** Used for process output data that list whether the product is defective or not. Examples include the p control chart, the np control chart, the c control chart and the u control chart. Unlike attribute control charts that count defective outputs, variable control charts are based on a quality output that can be measured numerically. Because this quality output is generally more informative than a pass/fail output description, variable control charts are preferred to attribute control charts when they can be used correctly.

**Attribute Data.** A quality characteristic that is classified into either conforming or nonconforming according to its specification is called an attribute. Attribute data are qualitative data that can be counted for recording and analysis. In general, this is lower quality data, since variable data contains potentially more information than just the attributes of the output.

**Autocorrelation.** If the measured output of a process steadily drifts away from its target value as the process goes out of control, the problem might be due to a slow but menacing accumulation or bad inputs. Autocorrelation refers to how items produced in time are related (or correlated) with each other due to this accumulation of error. Correlation is a more general term describing the relationship between process outputs, but autocorrelation specifically refers to observing this relationship between sequential measurements in time. It can also refer to spatial correlation (outputs close to each other in space).

**Average Run Length.** The number of points plotted on a control chart before it indicates an out-of-control condition is called the run length. The average number of run lengths you observe from a process is called the average run length and is used to evaluate the performance of control charts - the longer the average run length, the better the process.

**Bar Chart.** A simple chart of categories on the horizontal axis versus the category frequency on the vertical axis. This is also called a bar graph. A histogram is a special type of bar chart in which the horizontal axis contains the ordered values of a random variable, and the vertical axis shows the relative frequency.

**Bell Curve -** see Normal Distribution.

**Bias.** The difference between the expected value (or mean) of an estimator and the characteristic it is estimating is called bias. If the expected value is the characteristic, the estimator is unbiased.

**Binomial Distribution.** This is a special distribution of discrete data for the number of successes (0, 1, 2, , n) in n independent trials, where n is a positive integer. Here success and failure are arbitrary, but it is assumed that the probability of success remains the same for each of the n trials. The binomial distribution is usually used to characterize the number of failures in a production lot.

**Bivariate.** Data that have two numerical components are called bivariate and can be graphed as a scatter plot with a horizontal axis for measuring one component and a vertical component to measure the other. Bivariate data represent a special case of data with two or more components, which are called multivariate.

**Blocking -** This is a standard technique in experimental design for reducing the effect of an environment or some other factor of less importance on the response. Blocking does this by ensuring that all treatment levels are used in every different environment, so the effects of the environment are "blocked out" of our assessment of the natural variability in the output.

**Box and Whiskers Plot -** see Box Plot.

**Box Plot.** A special statistical summary illustration of a set of data that includes five fundamental statistics: the sample median, the upper and lower quartiles, and the smallest and largest value. The plot gets its name because an actual box is graphed to show the range between the lower quartile (or 25th percentile), the median, and the upper quartile (or 75th percentile). In some box plots, additional statistics are plotted, such as the sample mean.

**Brainstorming.** This is a collaborative technique for generating new and constructive ideas via teamwork. Brainstorming can be helpful in both finding assignable causes in a process and creating innovative ways of removing those problems.

**c Control Chart.** The c Control chart, or Counts Control Chart, is used to chart the number of defectives in cases where the number of defects in a lot is not limited to a fixed sample size. The Poisson distribution is used to describe this kind of count data in which we are counting the frequency of rare occurrences (defects) that have a nearly uncountable number of opportunities to happen. In cases where the sample size is known (e.g., counting each product as defective or not), the np control chart should be used.

**Cause and Effect Diagram.** This is a graphical tool in which potential error causes are listed on lines which branch off from their effects. Because it looks a bit like a fish skeleton, it is sometimes called a fishbone diagram.

**Central Limit Theorem.** A fundamental statistical result that says the distribution of the sample average will appear bell-shaped once the sample size is sufficiently large. Note that this is true even if the distribution of the original measurements do not look at all bell-shaped. The distribution of the sample mean is converging to a statistical distribution called the Normal distribution. The central limit theorem is important to statisticians because while the exact distribution of the sample mean might be impossible to compute, the normal approximation is relatively easy.

**Central Tendency.** This is a term that most generally describes the middle of a data set. Because there is no unique definition of central tendency, we use both the sample mean and the sample median to characterize it, even though they are often noticeably different in many samples. Other measures of central tendency include the mode and the trimmed mean.

**Chi Square Distribution.** This is one of the most important statistical distributions in process control because it often describes the distribution of the sample variance. The chi square distribution is used, for example, to construct the control limits for the s chart.

**Coefficient of Determination.** A useful statistic provided by analysis of variance or regression to describe how well the chosen model fits the data. For example, in a linear regression, the coefficient of determination represents the proportion of variability in the response that can be explained by the explanatory variables. In a simple linear regression, the coefficient of determination is the squared value of the correlation coefficient between the response and the regressor.

**Coefficient of Variation.** A measure of the data set's spread, defined as 100 times the standard deviation and divided by the mean. The coefficient of variation (CV) is useful for comparing variability in different populations, assuming these values are strictly positive. By dividing by the mean, the CV does not have units, and it provides a way of checking variability in the context of the population mean.

**Common Cause Variation** - see Noise.

**Confidence Interval.** For an unknown characteristic (like the mean or standard deviation), this is an interval of numbers that describes a set of plausible values for the characteristic of interest. The interval has a confidence associated with the interval in terms of a probability. For example, if a 90% confidence interval for the mean is (100, 105), then we are 90% certain the unknown population mean is somewhere between 100 and 105.

**Conformance -** see 8 Quality Dimensions.

**Conforming.** Simply stated, a conforming product is one that meets the required specifications. Note that a conforming product is not necessarily the same as a non-defective product. If detected product defects do not interfere with the product achieving the required specification, the product is still conforming.

**Confounding.** In some experiments, if two factors affect the product output, their effects will be confounded if their levels change together for every experimental output. For example, suppose factor A is set at two levels (low, high) and B is set at two levels (low, high). If (A, B) are both set to low levels for one experiment and both set to high for another, it will be impossible to tell which factor was the cause of the changing experimental outcome. Continuous Process Improvement. More than a slogan, in statistical process control, continuous process improvement (CPI) is described specifically in Section 4 of Chapter 1 through eleven steps: (1) defining the problem, (2) prioritizing problem areas, (3) organizing a CPI team, (4) documenting the process, (5) collecting data, (6) analyzing data, (7) developing improvement goals, (8) taking improvement action, (9) evaluating results, (10) standardizing improvement actions, and (11) recording the lessons learned.

**Control Chart.** The general term for graphing output levels of the process continually. Every control chart has a target value for the measured output and control limits to determine when the process is out of control. Some charts also have warning limits that let the process manager know the output is close to being out of control. Some charts require sample statistics such as the sample average or standard deviation, but some charts are based on single observations. Product output can be based on output measurements or merely counting conforming products. Every control chart has a sensitivity to changes in process quality and a given propensity to declare a process out of control when it is not.

**Control Limits.** For a control chart, the control limits define an interval of product measurement or other process output for which the process is deemed out of control. The process remains in control as long as it stays above the Lower Control Limit and/or below the Upper Control Limit. For charts plotting the product mean, both upper and lower control limits are typically applied. For charts plotting sample variance, only an upper control limit would matter to the process manager. Once a control limit is exceeded, the process is considered out of control and the process manager uses information from the process data to find potential problems.

**Controllable Factors** - see Factors, Controllable.

**Correlation.** For data that have two components (see bivariate data), the components are correlated if they vary together in some pattern. In such a case, knowing the value of one component will affect the expected value of the other component. Correlation is measured in a data set using the correlation coefficient. If bivariate data are uncorrelated, then knowing the value of one component has no affect on our knowledge of the other one. If the variables are independent, they will naturally be uncorrelated.

**Cp** - see Process Capability Index.

**CUSUM Control Chart.** An advanced chart that considers not only the most recently measured output from the process, but the average of all recent process outputs. If the process exhibits a drift in terms of a slowly decreasing or increasing process mean, a regular Shewhart control chart might miss the subtlety of the change. The CUSUM is designed to be more sensitive to this kind of drift because as it accumulates past information, the changing process mean will be more obvious. As a tradeoff, the CUSUM is less sensitive than the Shewhart charts in detecting immediate and dramatic changes in the process output.

**CV Control Chart.** This chart effectively monitors systems for which the variance is a function of the mean by computing and plotting the coefficient of variation. In such a case, the R or s control charts might fail to detect when the process is out of control. In some manufacturing processes, for example, the variation of the output is expected to increase as the mean increases, and such a process should be monitored using a CV control chart.

**Data.** The general term to describe the measured output of the process. By treating the data set as an experimental outcome, we use statistical methods to analyze the numbers as if they could have been generated differently. Using this outlook, we treat them as random outcomes.

**Decision Interval.** In a CUSUM control chart, the decision interval is an adjustable parameter that affects the chart's control limits. As a rule-of-thumb, the decision interval is set around four or five standard deviations.

**Decision Rule.** A function of process output that determines a discrete action, such as declaring a process out of control or declaring the process in control.

**Defect.** A flaw or mistake on the measured process output. If the flaw causes the product to fail in meeting required specifications, the product will be deemed nonconforming. If detected defects do not interfere with the product achieving the required specification, the product can still be considered as conforming.

**Degradation** - see Process Degradation.

**Deming Cycle.** An iterative problem solving technique, also called PDCA ("Plan-Do-Check-Act") that consists of four repeated steps: (1) Plan - Design objectives for the process to achieve required results, (2) Do - Put the process into operation, (3) Check - Monitor the process by measuring process output (e.g., using a control chart), (4) Act - Make necessary repairs and actions to implement continuous process improvement.

**Design of Experiments.** Experimental set up in which a response variable is chosen for measurement along with a set of controllable variables that might affect the response. Changes in the response variable should help the experimenter decide whether a scientific conjecture is true or false. In SPC, the process output can act as a response, and purposeful changes in the controllable inputs can be studied in order to find optimal output (in terms of meeting specification).

**Distribution.** Used here as a general term to describe the location and spread of a set of data generated from an experiment. Some populations have special functions that characterize the distributions generated from them, and are referred to by name (Normal, Chi square, Poisson, Binomial, and so on).

**Distribution Free** - see Nonparametric.

**DMAIC.** A six-sigma methodology for improving a process in five steps: (1) Define process improvement goals. (2) Measure important aspects of the process. (3) Analyze output and establish whether the process is under control. Find all potential factors that affect product output. (4) Improve the process by fixing problems found in the process to prevent future problems. (5) Control the process using SPC techniques, making sure any deviations from target are corrected before they defects occur.

**Durability** - 8 see Quality Dimensions.

**Engineering Process Control.** Industrial process controls that mechanically respond to process variation by adjusting process inputs. Since changes occur in real time, process outputs are usually autocorrelated (dependent on previous measurements). For example, automatic temperature and humidity readings can initiate programmed changes in heating and air conditioning procedures in an industrial plant.

**Equipment Variation** - see Repeatability.

**Estimator.** Any function of the data that is used as a surrogate for an unknown population characteristic. The most obvious example is a sample mean, which is an estimator for the population mean.

**EWMA Control Chart.** An advanced chart that considers not only the most recently measured output from the process, but a weighted average of all the most recent process outputs. If the process exhibits a drift in terms of a slowly decreasing or increasing process mean, a regular Shewhart control chart might miss the subtlety of the change. Like the CUSUM control chart, the EWMA control chart is designed to be more sensitive to this kind of drift because as it accumulates past information, the changing process mean will be more obvious. Unlike the CUSUM, the EWMA is more influenced by more recent measurements. Compared to the Shewhart charts, the EWMA is less sensitive in detecting immediate and dramatic changes in the process output.

**Exponential Distribution.** This is a special distribution of product measurement data, and is often used to characterize time to failure. The distribution has the memoryless property, which means that if the time to failure of a component has an exponential distribution, then the conditional (or residual) life of a used component is equivalent to that of a new one.

**External Failure Cost** - see Quality Costs.

**Factorial Design.** A special kind of experimental design in which the inputs (or factors) have a fixed number of possible levels (usually 2 or 3), and the experiment is carried out at some or all of the various combinations of input levels. For example, if a response is a function of three input factors, and each factor has two levels, then there are $2 \times 2 \times 2 = 8$ different combinations of inputs that are possible. This example is called a complete factorial design.

**Factors, Controllable.** Input factors of a process that can be manipulated to change process output. In a manufacturing process, the amount of raw material used or the range of operator decisions are considered controllable factors. Factors that change but are beyond the process manager's ability to manipulate are called uncontrollable factors.

**Factors, Uncontrollable.** Inputs to a process that clearly affect the output but are not under control of the process manager. Unlike controllable factors, which are changed in order to optimize the process output, variability due to uncontrollable factors is generally a cause for defects or process errors.

**Failure Modes and Effects Analysis.** The analysis of potential failure modes in order to determine the severity of the failures' effect upon the process. FMEA is typically implemented early in the process development in order to save money and time by finding and removing process flaws early.

**False Alarm.** In statistical process control, nonstandard outputs can lead to the decision that the process is out of control when it actually was in control. A false alarm is also considered a Type I error, opposed to a Type II error, in which the process is out of control, but the SPC decision rule is that the process is in control.

**Features** - see 8 Quality Dimensions.

**Fishbone Diagram** - see Cause and Effect Diagram.

**FMEA** - see Failure Modes and Effects Analysis.

**Gage R&R Studies.** R & R stands for repeatability and reproducibility. Gauge R&R studies are based on repeating measurements for key output characteristics in order to find out if inconsistencies in measurements are too large to ignore. Differences might show a faulty input or an inconsistency in operations involving the input.

**Goodness of Fit.** This term describes statistical tests that determine if the distribution of data can be identified as one that is commonly used in SPC, such as the normal distribution. Goodness of Fit is usually implemented as a verification tool. When the test results suggest the data are not matched with the assumed distribution, the statistician will rely on nonparametric techniques to draw inference on the data.

**Grade.**  A category assigned to products or services having the same functional use but different technical characteristics. Grade is separate from quality, which categorizes products only on this characteristic. High grade products generally refer to items that promise more and better features.

**Histogram.**  A visual display to show how the data are distributed. The histogram appears as a bar chart. Each bar represents the frequency measurement (on the vertical axis) for the data in a specified range, which is ordered on the horizontal axis.

**In Control.**  The way to describe a process when the measurements of the quality characteristic are within the specification limits dictated by the target value and the allotted error limits.

**Independence.**  For data that have two components (see bivariate data), the components are independent if the way they are distributed are unrelated to each other. Knowing the value of one component in the pair does not provide any additional useful information about the value of the other component. Bivariate data that vary together in a related way are correlated.

**Individuals Control Chart** - see X-MR Control Chart.

**Interaction.**  For two input factors of a process, interaction describes how the affect of one of the factors on the output changes when the other factor is set at a different level. This means that the analyst cannot consider the factors individually (or marginally) when assessing their affect on the response. In a fast food restaurant, for example, consider the factor of the number of coffee dispensers in use (1 versus 2). The effect of this choice on the output (time until customer is served) might be dependent on another factor: the type of meal that is being served (breakfast or lunch).

**Internal Failure Cost** - see Quality Costs.

**Ishikawa Diagram** - see Cause and Effect Diagram.

**ISO/QS 9000 Standards.**  This set of standards for quality management is managed by the International Organization for Standardization (ISO) using certification and accreditation procedures. ISO 9000 requirements include monitoring processes, and have led to an increase in SPC awareness and implementation in various industries.

**Kurtosis.** As an alternative description of variability in a data set, higher kurtosis suggests that much of the variance in the data is due to infrequent extreme deviations. Histograms for data with higher kurtosis will show a more peaked distribution (one bar is likely to be much higher than the others).

**Lead Time** - see Process Lead Time.

**Lean Methodology.** Lean refers to methods for improving process quality and efficiency via the elimination of waste. Due to its industrial origin, it is sometimes referred to as Lean Manufacturing. Lean emphasizes process speed or flow, and optimization techniques are used to guide companies in dealing with time constraints.

**Lower Bound.** As a one-sided confidence interval, the lower bound dictates a number for which we have a degree of confidence is below the characteristic of study. For example, when the small quantiles of a distribution of inputs of an uncontrollable factor are consequential to a process (e.g., 0.05 quantile) a lower bound offers assurances that a process won't be negatively affected.

**Lower Control Limit** - see Control Limits.

**Main Effects** - In an analysis of variance, this refers to observable changes in the response that are due to changing the levels of a factor. A factor's main effects do not depend on what the other factors are doing to the experiment.

**Manhattan Diagram.** If the output of a process is plotted in real time, the control chart will draw a continuous line rather than the jagged lines we expect in a Shewhart Chart. If the process has discrete time points at which the measurement or sample average is updated, the chart will be plotted as a step function, taking the appearance of a city skyline.

**Margin of Error.** This statistic typically refers to the amount of sampling error in a survey results. In Section 4 of Chapter 3, the margin of error was paired with a confidence value, reflecting the certainty that the unknown population characteristic was within the stated margin of error.

**Mean.** The sample mean refers to the arithmetic average of all the measurements of the sample. It is an estimator for the population mean. Along with the median and mode, the mean is a measure of central tendency for a set of data.

**Measurement System Analysis.** An analysis for evaluating the process used in obtaining measurements to ensure the reliability of the process input or output data. The goal is to understand the repercussions of measurement error in statistical decision making.

**Median.** The sample median refers to the middle value in the sample. For example, if there are five measurements in the sample, the median ranks as 3rd largest out of five. If there is an even number in the sample (so no unique middle value exists) we usually average the two middle values. It is an estimator of the population median. Along with the mean and mode, the median is a measure of central tendency for a data set.

**Median Control Chart.** Similar to the - chart, the median control chart plots the sample median instead of the sample mean. This is used in cases for which the distribution of output is highly skewed, so the sample median is considered to be a better representative of a typical value from the sample, compared to the sample mean. Because the - chart assumes the distribution of the sample mean is normal (or bell-shaped), the median control chart is considered more robust because it performs well even if this is not the case.

**Mode.** The sample mode is the value that occurs most frequently in the sample. The mode is used as a measure of central tendency only in cases where the process output is limited to a discrete set of values. In many processes where exact measurements are taken carefully, none of the observations are expected to have the same measured value, in which case the sample mean or sample median should be used instead of the sample mode.

**Monitoring a Process.** This is the general term for examining process outputs to ensure the process is in control. Statistical process control is a primary method for monitoring a process. This can also stipulate that the process manager keep an eye on process inputs, including both controllable and uncontrollable factors.

**Moving Range.** The moving range is a list of calculated differences between successive outputs of a process. The moving range chart plots the measured values of the process across time (also called an individuals chart). Unlike the s - chart or R - chart, the moving average chart is based on a single measurement per measuring time. For control limits, we use the absolute value of the successive differences, which is the moving range.

**Multiple Regression.** The statistical procedure for estimating an experimental response with a combination of input variables, or explanatory
variables. Multiple Linear Regression pertains to responses that can be
described as a linear combination of input variables. For example, the
amount of sewage (Y) produced by a manufacturing plant might be described as a linear function of the number of man-hours used by plant
operators (x1) and amount of raw materials arriving to the plant in shipment (x2). A regression will estimate this linear relationship by estimating
coefficients a, b and c so that Y is predicted by a + b x1 + c x2. Simple
Linear Regression represents a special case where only one input is used
to describe the response.

**Multivariate.** This refers to a measurement that has more than one component. For example, the quality of a process output might be measured
by the product weight, height and length. Instead of a single value to determine output conformity, this process has three, so the process output
is multivariate.

**Natural Variability** - see Factors, Uncontrollable.

**Noise** - see Factors, Uncontrollable.

**Non Normality.** This describes the distribution of data that does not conveniently adhere to a bell-shaped curve. Data anomalies such as skew or
the existence of outliers are a cause for non normality. This property of
the data can be illustrated using a normal probability plot.

**Nonconforming.** A nonconforming product is one that fails to meets the all
of the required specifications. A nonconforming product is not necessarily
the same as a defective product. If detected product defects do not interfere with the product achieving the required specification, the product is
still conforming.

**Nonparametric.** Describes a statistical analysis that does not presume the
data are distributed in a familiar manner, such as with a normal distribution. Unlike these regular statistical methods, nonparametric methods
perform well in all sorts of situations, but are sometimes less efficient
in cases where the distribution assumptions are met. While these other
parametric procedures often compare values and averages of data sets,
nonparametric procedures can be based on the order (or rank) of the
values and averages.

**Nonparametric Control Chart.**  In cases where the distribution is far from mound-shaped (like the bell curve), Shewhart control charts are susceptible to inaccuracies regarding control limits. Although Shewhart charts are generally robust, if the data are highly skewed or otherwise non normal, nonparametric control charts can be used to monitor the process. Unlike typical charts, the nonparametric control chart will identify an out of control process by how the observations are ranked or categorized across time. For example, if a process produces an output above its target mean five times in a row, then assuming this occurs half the time, the nonparametric procedure is to estimate it happens five times in a row only once in $2 \times 2 \times 2 \times 2 \times 2 = 32$ trials. If $1/32 = 0.0313$ is low enough to trigger an out-of-control signal, the process will be stopped to see if there is a bias or flaw in the system.

**Normal Distribution.**  This describes the distribution of measurements that are symmetric about an average, giving the corresponding histogram the look of a "bell-shaped curve". While very few samples actually look exactly like they were from this distribution, it nonetheless provides an adequate approximation to the distribution of many types of data. The central limit theorem in Section 3.3 provides further motivation for assuming the distribution of process output measurements are "normal".

**np Control Chart.**  For plotting attribute data (outputs are classified only as defective or non defective), the np control chart plots the number of nonconforming products. Since this could be misleading if the sample or lot size changes at different inspections, it is assumed the lot sizes are equal for the np control chart. If sample sizes are changing, the p control chart should be used instead.

**Null Hypothesis.**  In a statistical decision making process, conjectures about the true nature of the process are split into two possibilities: the null hypothesis and the alternative hypothesis (see Section 3.5). When claims are made about process outcomes, the null hypothesis usually coincides with the claim that the measurement variability is due to randomness or noise. The alternative hypothesis, on the other hand, goes with the claim that changes in process output are due to assignable causes.

**Operator Variation** - see Reproducibility.

**Outlier.**  While most observations from a set of measurements are clustered around some sample mean, an outlier represents an observation that seems as if it doesn't belong with the rest. In some cases, an outlier could give the process manager insight on process capabilities and potential problems. In other cases, the outlier might be only represent a typo during the data entry process.

**Out-of-control** - see In control.

**Out-of-spec.** Technical jargon for describing products that fail to meet specification limits.

**p Control Chart.** For plotting attribute data (outputs are classified only as defective versus non defective or conforming versus non conforming), the p control chart plots the fraction of nonconforming products. If the sample size changes greatly between inspection times, the chart's control limits will also. With larger sample sizes, the p chart will have narrower control limits, reflecting more certainty in the estimate of the true fraction of nonconforming products that are produced by the process. If sample sizes are constant, the np control chart can be used.

**Pareto Chart.** This charts frequency of categorized outputs in order of descending frequency. That is, the most frequently observed category is plotted first (on the left), and the least frequently one is last (on the right). If process defects are categorized, this plot exploits the Pareto principle by showing a sharp decline in category frequency after the first or second category, which suggests that most of the defects are due to one or two causes.

**Pareto Principle.** This rule applies to processes for which 20% of the identified problems cause 80% of the process damage. The principle is named after Vilfredo Pareto, who first hypothesized that 80% of Italy's land (in 1896) was owned by only 20% of the population. It is generally accepted that the Pareto principle can be extended to other "symmetric" percentages (10% vs. 90%, for example).

**PDCA** - see Deming Cycle.

**Perceived Quality** - see 8 Quality Dimensions.

**Percentiles.** In our context, percentiles refer to the ranked observations of a sample of data. For example, the median, or middle value of the sample, is called the 50th percentile. The 90th percentile would be the value for which 90% of the measurements in the sample are less, and 10% of the measurements are more.

**Performance** - see 8 Quality Dimensions.

**PERT Analysis.** PERT is an acronym for Program, Evaluation, and Review Technique. Similar to a critical path analysis, the PERT analysis uses statistical analysis to determine optimal times, the most probable times, and the least optimal times for completion of each activity.

**Phase I Monitoring.** If the target (mean) value or the natural variability of the process output is unknown, an initial process run is used to gauge this using the sample mean and sample variance. This stage is useful, of course, only if the process is known to be in control.

**Phase II Monitoring.** This is the name for the main phase of statistical process control, when target values for process output (including mean and/or variability) are used to construct control limits for a control chart. Monitoring uses these charts to detect shifts or excessive variability in the process output, at which time the process manager looks for assignable causes for this change.

**Pie Chart.** If product defects are classified into discrete categories, their relative frequencies can be represented in a circular graph as different slices of pie. For example, if a category frequency is 25 percent of the data set, then the slice representing that category will take up one-fourth of the pie. The pie chart has an advantage over the bar chart for examples in which frequency doesn't matter as much as relative frequency; that is, the frequency found in each category matters only with respect to frequencies found in the other categories.

**Poisson Distribution.** This distribution is used when we are counting the frequency of rare or uncommon occurrences (product defects) that might have a nearly uncountable number of chances to occur. It is for discrete data, like the Binomial distribution, but not for counting the number of defective products from of a fixed lot size. For example, if an inspector counts the number of flaws on the final paint job for a new automobile, we might expect a mean count of one or two, but the Poisson distribution is used for these kind of counts because we could conceptually find an unbounded number of flaws on the paint surface of the auto.

**Power.** In our context, power refers to the probability a statistical test can find and conclude a conjectured effect when the effect is waiting to be found. Because type II error is the chance of not finding the effect when it is present, it complements statistical power. In terms of statistical hypotheses, the effect we are conjecturing is synonymous with the alternative hypothesis.

**Prediction Interval.** For a future observation from the process output, this is an interval of numbers that describes a set of plausible values for the unknown output value. The interval has a confidence associated with the interval in terms of a probability, and is based not only on our uncertainty about this future value, but also the uncertainty associated with the statistical method used to make the prediction. For example if a 90% prediction interval is (100, 105), then we are 90% certain the future observation is somewhere between 100 and 105.

**Prevention Cost** - see Quality Costs.

**Probability Plot.** If a set of data are distributed normal (or approximately so), then the transformation of points that makes up a (normal) probability plot will form a straight line. If the data are distributed in some other way, the plot will reveal some kind of curvature that is detectable to the observer. These plots are commonly used to judge whether the data are normally distributed or not, because many procedures in statistical process control assume the measurements have this property. There are plots for other kinds of distributions, too, that can be accomplished with a probability plot. For that reason, this is more specifically called a *Normal Probability Plot*.

**Probability Limits** - see Control Limits.

**Process.** The primary activity that turns inputs (e.g., raw materials, labor) into outputs (e.g., products, service).

**Process Capability Analysis.** This analysis consists of evaluating the performance of a process outputs in terms of its specifications. The process is considered capable if all of the process output values fall within the specification limits.

**Process Capability Index.** This is an indicator for whether a process can meet requirements imposed by the process constraints. These constraints might be internal (physical limits of the production process) or external (viability for customer appeal). The indicator (Cp) is based on the given control limits and variability for an in-control process. This is often called a process capability ratio because the index value is based on the ratio of target capability and observed capability. For example, a common index of $Cp = (UCL - LCL)/6\sigma$, the $6\sigma$ spread of the process is the process capability (for normally distributed data).

**Process Degradation.** Degradation refers to measurable decline of process capability through shifting product output or variability. Some products that degrade will fail at an increasing rate, such as a capacitor. Other products, such as an automobile tire, are taken out of use when the measured degradation reaches a fixed threshold value.

**Process Lead Time.** The average time it takes for one unit to go through the entire process, computed as the ratio between the number of items in the process and the number of items in the process completed per hour.

**Process Shift.** An upward or downward change in the center of variation of a previously stable process.

**Proportion Defective.** The ratio of defective products in the inspected lot to the lot sample size. The p - control chart plots the proportion defective over sequences of lot inspections.

**p - Value.** For a given sample (or sample test statistic) and a test of hypothesis in statistics, the p-value represents the probability of generating a sample that seems as consistent with the alternative hypothesis given the null hypothesis is true. For example, suppose a sample produces a test statistic (e.g., sample mean) that corresponds to a p-value of 0.05. This means that if we generated simulated data from the distribution described in the null hypothesis, only 5% of the time would the test statistic be more in agreement with the alternative hypothesis. For this reason, we tend to reject the null hypothesis if the p-value of the test statistic is small, say smaller than 0.05.

**Qualitative Data.** A type of data that cannot be ordered and compared like numbers. Instead, qualitative data must rely on categories that uniquely partition the data set.

**Quality.** The degree to which the product meets customer specifications. The ISO 9000 defines quality as the "Degree to which a set of inherent characteristic fulfills requirements." Quality Control. A collection of methods used to ensure a process produces outputs that meet customer requirements. For typical processes in manufacturing and the service industry, statistical quality control is the chief way of meeting these needs during the production process.

**Quality Costs.** The cost exacted on the producer for doing things wrong. Section 1.5 lists four categories of quality costs: prevention cost, appraisal cost, internal failure cost, and external failure cost.

**Quality Management.** This is a general method for ensuring that the steps needed to design, develop and implement a product or service are proven to be effective and efficient. Quality control is a primary component for quality management.

**Quantiles.** These are numbers (proportions) between zero and one that correspond to the ranked observations of a sample of data. For example, the median, or middle value of the sample, is called the 0.5 quantile. The 0.9 quantile would be the value for which 90% of the measurements in the sample are less, and 10% of the measurements are more.

**Quantitative Data.** Numeric data (not necessarily integers) which can be ordered, compared and are subject to mathematical functions (like adding or subtracting).

**Quartile.** There are three quartiles in a set of data, corresponding to the 25th 50th and 75th percentiles in the data set. In this way, the quartiles partition the data set into four ordered and equally sized parts. The quartiles can be illustrated using a Box Plot.

**R Control Chart -** see $\bar{x}$ - R Control Charts.

**Radar Chart.** A data display with several axes, each representing one of the categories found in the data. The axes protrude outward from the center according to the frequency of observations in that category. The radar chart is illustrated in Section 2.6.

**Range.** Do describe dispersion in a data set, the range is defined to be the distance between the smallest observation in the data and the largest one. When sample sizes are small, say less than ten, the range is sometimes preferred over the sample variance.

**Reference Value.** In a CUSUM control chart, the reference value is an adjustable parameter, chosen to be one half of the magnitude of shift needed to detect. If shift is expressed in standard deviations, then the reference value can be assigned as $\sigma/2$.

**Regression -** See Multiple Regression.

**Regression to the Mean.** This refers to the failure of taking natural variation into account when ascribing causes to outliers. It occurs when an extreme measurement is observed and followed by a more expected outcome that is closer to the mean.

**Relative Precision Index.** To gage process capability, the relative precision index (RPI) is calculated as (UCL – LCL)/R, where UCL is the upper control limit, LCL is the lower control limit, and R is the average value of the sample ranges used in the chart. The RPI has a role similar to the process capability index.

**Repeatability.** In measurement analysis, this represents the variation in measurements obtained by a single gage when used several times by the same operator on the same product.

**Reproducibility.** In measurement analysis, this is the variation in the average of the measurements made by different operators.

**Residuals.** In statistics, residual refers to the difference between an actual sample measurement and what the predicted value of that measurement would be using statistical estimation. In regression, for example, the residual is the distance between the response and the value of the regression line, given the same value of regressors.

**Robust.** This describes any statistical method that tends to be effective even if the assumptions of the data turn out not to be true. For example, some statistical procedures assume the underlying data have a normal distribution. A subset of these techniques will give misleading answers, but robust techniques will be more valid.

**Run Length.** In a control chart, the number of points plotted before it indicates an out-of-control condition is called the run length. The average run length (ARL) is the expected run length given the process is in control.

**s Control Chart -** see $\bar{x}$ - s Control Charts.

**Sample Size.** The number of unique process outputs that are measured at each time point between sampling intervals. For attribute data (defect or non defect), sample size refers to the number of inspected products in the lot.

**Sampling Interval.** The length (in time or inspected units) between sequential sampling in process output. In a continuous process, items may be sampled once for every 100 units manufactured, for example. Alternatively, the process could be stopped at fixed time intervals in order to sample the most recent process outputs.

**Saturated Design.** In a design of experiment, if the relationship between the inputs and outputs are too complicated, there may be more things to estimate than there are data. Cases in which there is not enough information left over to construct a valid estimator of sample variance are called saturated designs.

**Scatter Plot.** For bivariate data, a scatter plot represents one component on the horizontal axis, and the other on the vertical axis. These plots are used to identify relationships between the two components from a sample of bivariate data.

**Serviceability -** see 8 Quality Dimensions.

**Shewhart Control Chart.** For a process output that has a target mean and natural variation defined by its standard deviation $\sigma$, the Shewhart control chart graphs process output values and declares the process out of control if the measurements are $3\sigma$ or more away from the target mean. Some charts also contain warning limits $2\sigma$ away from the target mean to let the process manager know the output is close to being out of control. Some Shewhart charts require sample statistics such as sample average or standard deviation, but some charts are based on single observations. Product output can be based on output measurements or merely counting conforming products.

**Simple Linear Regression -** see Multiple Regression.

**Six Sigma Process Quality.** This refers to a management approach that seeks to identify and eliminate causes of defects in a process. Six sigma relies heavily on statistical methods, especially statistical process control. Six sigma strategies originated in Motorola's manufacturing plant, but have been extended to all types of businesses and industries.

**Skew.** In a sample of data, skew refers to distributions in which the relative frequency appears to increase or decrease on one side, making the distribution look asymmetric. Illustrated by histograms in Section 2.7, skewed distributions sometimes represent distributions of positive measurements, especially when most measurements are small (near some nominal middle value) but a minority of observations tend to be much larger (two or three times, perhaps) than the central value of the distribution.

**Special Cause Variation -** see Assignable Cause Variation

**Spider Chart -** see Radar Chart.

**Stability.** In process monitoring, a stable process is one that is deemed in control. In measurement analysis, stability refers to the variability obtained with a gage on the same product over an extended period of time.

**Standard Deviation.** This is the primary statistic used to represent dispersion and item-to-item variability within a sample. Calculated in Section 3.2, the sample standard deviation (s) is based on squared differences between sample units and the sample mean, so if all of the sample items are clustered together (near the sample mean), then the standard deviation will be small. It is based on a square root of these squared errors so the units of standard deviation are the same as the original measurements. The term "standard deviation" can also refer to the unknown variability in the population (denoted $\sigma$) which is estimated by the sample standard deviation, denoted by the letter $s$.

**Standard Error.** This usually refers to the calculated standard deviation of a sample mean. In this case, standard error is smaller than standard deviation because the sample mean is based on more measurements than a single sample observation. More specifically, the ratio of the standard deviation to the standard error is the square root of the sample size. For example, if the sample size is 100, the standard error is 10 times smaller than the standard deviation.

**Statistical Process Control.** The use of statistical methods in process control and process improvement.

**t - test.** A statistical test for an unknown population mean based on one or more sample means. A key property of the t-test is that the population variance is unknown and estimated by the sample variance. Any sample mean used in a t-test is assumed to have a normal, or bell-shaped, distribution.

**Tabular CUSUM.** A method for creating control limits for the CUSUM control chart based on counting process outputs that are above the target and process outputs that are below the target. These statistics, listed in Section 5.1 are used along with two parameters, called the reference value and the decision interval, to construct the CUSUM chart.

**Taguchi Method.** This method centers on design of statistical experiments, stressing designs that include an effect of variance. To estimate variance components, Taguchi promoted the idea of replicating the experiment at the given settings. Tally Charts. The name of our most primitive charting method - adding a notch or a mark besides a category name to keep track of its frequency.

**Target value.** The intended value of the measured quality characteristic for an in-control process.

**Test of Hypothesis.** The statistical framework adapted to make yes/no decisions about unknown facts and parameters about the process output. In the decision making process, conjectures about the true nature of the process are split into two possibilities: the null hypothesis and the alternative hypothesis (see Section 3.5). From the data, a test statistic is formulated to make a decision rule about the hypotheses.

**Total Quality Management.** TQM is a management policy for promoting awareness of quality in all organizational processes. Statistical process control is a chief feature of total quality management. Troubleshooting. This is sometimes vaguely defined in industry practice, but generally implies a systematic investigation for a problem source, usually by following the steps of the process from start to finish. Troubleshooting is often a process of elimination based on a careful scrutiny of each critical component of the system or process.

**Type I Error.** In a test of hypothesis, there are two possible mistakes that can be made. The type I error represents the mistake of rejecting the null hypothesis when it was actually true. In terms of statistical process control, the null hypothesis might mean the process is in control, and the alternative hypothesis implies the process is out of control. By making a type I error, we incur the cost of stopping the process for troubleshooting and unnecessary repairs when they were not needed.

**Type II Error.** Along with the type I error, these are the two mistakes that can be made in a test of hypothesis. The type II error represents the mistake of not rejecting the null hypothesis when it was actually false. In terms of statistical process control, the null hypothesis might mean the process is in control, and the alternative hypothesis implies the process is out of control. By making a type II error, the process is out of control, but we judged it to be in control, thus allowing the production of defective outputs from the process to continue. The power of a statistical test is the complement of type II error: power = 1 - P(type II error).

**u Control Chart.** Similar to the c -Control chart, the u - chart is used to chart the number of defectives in cases where the number of defects in a lot is not limited to a fixed sample size. The Poisson distribution is used to describe this kind of count data in which were are counting the frequency of rare occurrences (defects) that have a nearly uncountable number of opportunities to happen. In cases when a different number of units (or different unit sizes) appear at different inspections, the u - control chart should be used instead of the c - control chart because it based on the rate of defects per sample (or size unit), making the sample-to-sample comparisons more fair.

**Uncontrollable Factors** - see Factors, Uncontrollable.

**Upper Bound.** As a one-sided confidence interval, the upper bound dictates a number for which we have a degree of confidence is above the characteristic of study. For example, if we are recording the number of defects on a process output, a 90% upper bound gives us a number for which we can safely assume, with 90% certainty, the true mean number of defects produced must be lower than.

**Upper Control Limit** - see Control Limits.

**Variable Control Chart.** A Shewhart-type control chart that are used to evaluate variation in a process where the measurement is a variable. Typically, central tendency ($\bar{x}$ - chart) and dispersion (s - chart, r - chart) are monitored by the variable control chart.

**Variance.** Though the word is sometimes used to describe a general propensity of data dispersion, variance is technically defined in Section 3.2 as a statistic based on squared differences between sample units and the sample mean. This sample variance ($s^2$) is an estimator of the unknown variance in the process output ($\sigma^2$). The square root of the sample variance, called the standard deviation, is often preferred as a description of sample variability because units of the standard deviation are the same as the original units.

**V-Mask CUSUM.** Along with the more popular Tabular CUSUM, the V-Mask presents a way of generating control limits for the CUSUM chart. Using a complex computational method, the V-mask starts the control limits together at the most recent sample point, then they split (up and down) as you go back on the CUSUM chart, creating a visual V.

**Warning Limits.** Similar to control limits, warning limits define an interval of product measurement or other process output. If the plot goes beyond the control limits, the process is not yet considered out-of-control, but the process manager is signaled in case the process is degrading and on its way to being out-of-control. If upper and lower control limits are used, then upper and lower warning limits can be used simultaneously.

**Weighted Average.** When averaging measurements in a set of data, a sample average adds up the numbers and divides by the sample size. In some cases, certain observations might be considered more relevant than others, and will be divided by different numbers, and not necessarily the constant of sample size. If we write , the regular un-weighted mean assigns weights $w_i = 1/n$. As long as they add up to one (that is, $\sum_i w_i = 1$ ) the estimator of the mean is valid. x-MR Control Chart. This is a pairing of two Shewhart charts for processes that can produce only one process output value per sample. The x - chart is just the plot these individual values (this is sometimes called an individuals chart) in order to monitor the process mean and find out if a mean shift occurs. To monitor variability, the moving range (MR) chart computes a moving range, which is a list of calculated differences between successive outputs of a process. The MR chart plots the measured values of the process across time. For control limits, we use the absolute value of the successive differences, which is the moving range.

**$\bar{x}$ - R Control Chart.** This is the most commonly used pairing of Shewhart charts for monitoring the process mean and variability. The $\bar{x}$ chart plots sample means in order to detect shifts in the process mean. The R chart plots the sample range in order to monitor sample-to-sample variability. Usually, both upper and lower control limits are used on the $\bar{x}$ - chart, but only high variance is of concern for process output, so the R chart uses only an upper control limit.

**$\bar{x}$ - s Control Chart.** Similar to the $\bar{x}$ -R control chart, the $\bar{x}$ - s control charts are Shewhart charts for monitoring the process mean and variability. The $\bar{x}$ - chart plots sample means in order to detect shifts in the process mean. The s chart plots the sample standard deviation in order to monitor sample to sample variability. Usually, both upper and lower control limits are used on the $\bar{x}$ - chart, but only high variance is of concern for process output, so the s chart uses only an upper control limit.

**z Control Chart.** The z chart allows the process manager to monitor a system that produces outputs with a changing target value. Instead of plotting the output value (or sample mean), a standardized value z is computed by subtracting the known target value and the standard deviation. This implies both mean and variance must be known, even though they might be changing during the monitoring of the process.

# INDEX

Printed in the United States
By Bookmasters